Multilateral Environmental Agreements in Global Governance

This book proposes a conceptualisation of Multilateral Environmental Agreements as a form of International Organisations, exploring the ways in which they have expanded over time by discussing the nuances of authority in global governance.

Multilateral Environmental Agreements are the key type of cooperation between states to address environmental concerns globally. While their activities regularly attract much attention from an academic and non-academic audience, their peculiar hybrid nature in-between treaties and full-fledged International Organisations, means they are often underestimated. This book proposes a new and innovative conceptualisation of Multilateral Environmental Agreements as a specific type of International Organisation which allows for a more accurate understanding of their dynamic nature, and uncovers expansive tendencies which have so far gone almost completely unnoticed. Based on a modern understanding of authority in global governance, the book shows how Multilateral Environmental Agreements represent a separate entity, and expand beyond the boundaries originally set by their member states. The book draws upon the neo-functionalist concept of spillover, as well as multiple other theoretical frameworks, to identify the two main drivers of expansion in Multilateral Environmental Agreements. To illustrate these drivers, the empirical chapters conduct six structured case studies, analysing specific cases of authority expansion in ozone and climate protection, including the Green Climate Fund under the UN Framework Convention on Climate Change. Overall, this book offers an invaluable contribution to the theoretical discussion on informal types of organisations, and provides new and extensive empirical insights into unexpected past and recent developments in global environmental governance.

As such, this book will be of great interest to students and scholars of International Relations and International Law who focus on informal types of cooperation and legislation, dynamic institutional development, and international environmental politics generally.

Linda Maria Spielmann is a senior consultant at the Association of Cooperatives Bavaria (GVB) and works with cooperative banks on sustainability and other topics. Previously, she was a senior consultant with the ESG Advisory Financial Services team at BDO. Before that, she was a research associate and Lecturer in International Relations at University of Bamberg, Germany.

Routledge Research in Global Environmental Governance

Global environmental governance has been a prime concern of policy-makers since the United Nations Conference on the Human Environment in 1972. Yet, despite more than nine hundred multi-lateral environmental treaties coming into force over the past forty years and numerous public-private and private initiatives to mitigate global change, human-induced environmental degradation is reaching alarming levels. Scientists see compelling evidence that the entire earth system now operates well outside safe boundaries and at rates that accelerate. The urgent challenge from a social science perspective is how to organize the co-evolution of societies and their surrounding environment; in other words, how to develop effective and equitable governance solutions for today's global problems.

Against this background, the *Routledge Research in Global Environmental Governance* series delivers cutting-edge research on the most vibrant and relevant themes within the academic field of global environmental governance.

Series Editors:
Philipp Pattberg, VU University Amsterdam and the Amsterdam Global Change Institute (AGCI), the Netherlands.
Agni Kalfagianni, Utrecht University, the Netherlands.

Climate Change and the Endurance of Democracy
Daniel Lindvall

Multilateral Environmental Agreements in Global Governance
Organisational Dynamics and Authority Expansion
Linda Maria Spielmann

For more information about this series, please visit: www.routledge.com/Routledge-Research-in-Global-Environmental-Governance/book-series/RRGEG

Multilateral Environmental Agreements in Global Governance

Organisational Dynamics and Authority Expansion

Linda Maria Spielmann

First published 2025
by Routledge
4 Park Square, Milton Park, Abingdon, Oxon OX14 4RN

and by Routledge
605 Third Avenue, New York, NY 10158

Routledge is an imprint of the Taylor & Francis Group, an informa business

British Library Cataloguing-in-Publication Data
A catalogue record for this book is available from the British Library

ISBN: 9781032982229 (hbk)
ISBN: 9781032982236 (pbk)
ISBN: 9781003597681 (ebk)

DOI: 10.4324/9781003597681

Typeset in Sabon
by Newgen Publishing UK

Contents

Acknowledgements

The writing of this book would not have been possible without the intellectual, professional, and personal support of many others. Out of so many, a few nonetheless stand out.

I owe a special intellectual debt to Thomas Gehring, who drew my attention to the peculiar institutional features of Multilateral Environmental Agreements and allowed me to combine my personal passion for the environment with my research interests in international institutions. He shaped my approach in the most fundamental way and provided continuous support in navigating the process of bringing this book together. I also thank Monika Heupel for her excellent comments which helped to shape the final version of this book.

I am extremely grateful to the many friends and colleagues who were a part of this journey. Simon Linder discussed the details of informal institutions with me in depth and supported my case studies with his extensive knowledge on financial institutions and financial politics. Eva Stocker was an amazing friend throughout this whole process and would never let me give up. Thank you for your endless patience and support. Frank Bandau, Manuel Becker, Fabio Bothner, Claudia Genslein, Chris Kearney, Anja Lippstreu, Johanna Pschierer, Thomas Rixen, and Fabian Schmidt all provided valuable insights, encouragement, and support which shaped the text in various ways.

I further want to express gratitude to Rosie Anderson at Routledge who showed great interest and enthusiasm for my project and guided me through the publishing process. I thank the three reviewers who provided excellent comments which helped to sharpen my contribution. And, I thank all those at Routledge who were a part of the process from bringing this project from proposal to the final book.

The writing of this book has been supported by funding from the Heinrich Böll Foundation.

Linda Maria Spielmann
Munich, Germany
November 2024

Abbreviations

AFB	Adaptation Fund Board
CDM	Clean Development Mechanism
CEIT	Countries with Economies in Transition
CER	Certified Emission Reduction Credit
CFC	Chlorofluorocarbon
CITES	Convention on International Trade in Endangered Species
CMA	Conference of the Parties serving as the Meeting of the Parties to the Paris Agreement
CMP	Conference of the Parties serving as the Meeting of the Parties to the Kyoto Protocol
CO2	Carbon Dioxide
COP	Conference of Parties
EC	European Community
EEC	European Economic Community
EPA	US Environmental Protection Agency
ERT	Expert Review Team
EU	European Union
GATT	General Agreement on Tariffs and Trade
GCF	Green Climate Fund, Green Climate Fund
GEF	Global Environment Facility
GGDP	Global Governance Programme for Development
GHG	Greenhouse Gas
GNI	Gross National Income
GNP	Gross National Product
HFC	Hydrofluorocarbon
IEA	International Energy Agency
ILC	International Law Commission
ILO	International Labour Organisation
IO	International Organisation
IOE	Informal IO Empowerment
IPCC	Intergovernmental Panel on Climate Change
IR	International Relations

JI	Joint Implementation
JWG	Joint Working Group
LDCs	Least Developed Countries
MEA	Multilateral Environmental Agreement
MLF	Multilateral Fund of the Montreal Protocol
MOP	Meeting of the Parties to the Montreal Protocol
NAMA	Nationally Appropriate Mitigation Actions
NAPA	National Adaptation Plans of Action
NDC	Nationally Determined Contribution, Nationally Determined Contribution
NGO	Non-Governmental Organisation
OECD	Organisation for Economic Co-operation and Development
SBI	Subsidiary Body for Implementation
SBSTA	Subsidiary Body for Scientific and Technological Advice
SIDs	Small Island Developing States
UNCTAD	United Nations Conference on Trade and Development
UNDP	United Nations Development Programme
UNEP	United Nations Environment Programme
UNESCO	United Nations Educational, Scientific and Cultural Organisation
UNFCCC	United Nations Framework Convention on Climate Change
WMO	World Meteorological Organisation

Introduction

Multilateral Environmental Agreements (MEAs) are a peculiar form of cooperation that has become the preferred method for states to address environmental concerns globally. More than mere treaties, they possess institutional structures which continuously carve out the specifics of their own terms, but less than traditional International Organisations (IOs), these structures do not amount to the same large-scale bureaucracies. Their true nature is almost impossible to grasp with established approaches to the study of International Relations (IR). While they are commonly understood as empty shells structuring state interaction, there is strong reason to believe that this severely underestimates their true capacities. Many cases can be observed in which MEAs take on a life of their own, expanding their authority without formal amendments ratified by their member states but merely through adoption by their main decision-making body, typically a Conference of the Parties (COP). Scholars of International Law have begun to recognise that MEAs are "virtually identical to international organisations" (Brunnée 2002, p. 16). More recent approaches to the study of authority in global governance provide a starting point to move toward a similar understanding from an IR perspective (see for example, Gehring and Spielmann 2023). Since IOs are generally accepted to be separate entities with a tendency to expand, undertaking this step moves into focus the study of expansive tendencies of MEAs. However, because of their more ad hoc nature, established approaches to the study of IOs cannot be transferred directly. An investigation into the MEA-specific driving forces of expansion becomes necessary. Why do MEAs expand their authority beyond their original boundaries without formal delegation?

The investigation undertaken in the following chapters will try to find an answer to this question. I will proceed in three steps, beginning with the establishment of an understanding of MEAs as a specific type of organisation exercising authority over their member states. Then, I derive hypotheses about specific driving forces of MEA expansion by combining recent approaches to the study of authority with neofunctionalism as one of the most prominent theories of institutional expansion (usually reserved for the European context) in a new and innovative approach. Finally, I conduct a

DOI: 10.4324/9781003597681-1

number of structured case studies to uncover the main driving force of six specific instances of MEA authority expansion in the areas of ozone and climate protection.

I conclude that informal authority expansion in MEAs occurs where gaps and ambiguities in the existing institutional framework become relevant and the ratification process necessary for formal treaty amendments would pose a significant obstacle to resolving the issue. The central driving force of informal MEA authority expansion is functional pressure resulting from gaps and ambiguities in their typically skeletal founding treaties. This is, however, not the only way expansions become possible. As will be shown, under suitable circumstances, actors inside an MEA can also strategically employ its structures to realise an expansion that is not functionally necessary but benefits them beyond the functionality of the MEA. What all cases seem to have in common from an empirical standpoint is that they share their roots in the distinction between developing and industrialised country parties, between two groups with widely differing capacities when it comes to protecting the environment. These insights significantly advance our understanding of informal institutional dynamics in environmental politics and beyond, and will be of interest not only to those investigating MEAs specifically but also those interested in new forms of authority on the international level and other forms of informal governance and institutional adaptation.

To set the stage, this introductory chapter will provide a more detailed overview of the emergence of MEAs as the dominant form of cooperation in environmental politics and their specific structures, which are unusual in other areas of international politics. Further, it will carve out the question of the driving forces of their expansive tendencies in more detail, and provide an overview over the specific steps that will be undertaken to answer this question.

Defining Multilateral Environmental Agreements

It was not until the 1970s that mankind began to understand the burdens of an ever-expanding web of natural resource use and abuse placed on generations to follow, and fears of incalculable and unconscionable consequences began to invade our minds. This was also when environmental concerns began to feature more heavily on the international agenda. Since then, International Law has developed in leaps and bounds in the service of the environment. In more recent decades, climate change has risen to the top of the list of concerns not only of our politicians but of many people in their daily lives and considerations of their own personal futures. It is understood now that while the cost of compliance with environmental agreements may well be high, the cost of non-compliance will be even higher (Rajamani 2012, p. 605). And, with this understanding comes the unfathomable challenge of governing the rescue of our planet through international politics.

Even though proposals for a "World Environmental Organisation" as a common gravity centre of international environmental policy-making have been made, this idea has not gained momentum (Oberthür and Gehring 2004). Overcoming the sporadic and ad hoc nature of earlier attempts, international environmental regulation today takes place through a considerable number of specialised independent institutional arrangements, many of which operate under the umbrella of the UN. These treaties cover areas of common concern and create institutions for continuing development, implementation, and review. They tailor commitments to capacities through differential treatment of developed and developing countries and attract near-universal participation (Rajamani 2012, p. 610). The origins of this approach can be traced back to the United Nations Conference on the Human Environment, which took place in Stockholm in 1972. From this point on, numerous MEAs have been adopted, among them, the UNESCO Convention Concerning the Protection of the World Cultural and Natural Heritage (1972), the Convention on International Trade in Endangered Species (CITES) (1973), the Convention on Migratory Species of Wild Animals (1979), the Vienna Convention on the Protection of the Ozone Layer (1985) supplemented by the Montreal Protocol on Substances that Deplete the Ozone Layer (1987), and the Basel Convention on the Transboundary Movements of Hazardous Waste (1989). The Rio Earth Summit of 1992 provided further impetus for the negotiation and adoption of a number of further MEAs. These include most prominently the UN Framework Convention on Climate Change (UNFCCC) (1992) supplemented by the Kyoto Protocol (1997) and later the Paris Agreement (2015) but also the UN Framework Convention on Biological Diversity, the Convention to Combat Desertification (1994), the Rotterdam Convention on Prior Informed Consent (1998), the Cartagena Protocol on Biosafety (2000), and the Stockholm Convention on Persistent Organic Pollutants (2001).

While addressing a broad spectrum of various environmental concerns, these agreements establish a common pattern, which has remained specific to the realm of environmental regulation. Most MEAs consist of a skeletal framework convention, often supplemented by protocols or annexes, which include a number of selective authorisations for secondary decisions to be taken further down the line. For this purpose, MEAs usually comprise a number of treaty bodies organised around a Conference of Parties (COP). As an MEAs plenary organ, the COP usually consists of high-level representatives of all member states, meets regularly (usually annually), and serves as the supreme decision-making body of its constitutive agreement. Specific powers and procedures may vary, but decisions made at COP level become binding for all members (Churchill and Ulfstein 2000, pp. 623–626).

COPs enjoy significant power at the internal and external level and fulfil a variety of functions. Internally, the treaty explicitly mandates the COP to adopt rules of procedures, financial regulations, and the budget, establish subsidiary bodies and provide guidance to these bodies as well as the secretariat. Externally, COPs of many MEAs are authorised to make decisions relating to state parties' substantive obligations and supervise parties' implementation of and compliance with the MEA, as well as decide on the consequences of non-compliance (Churchill and Ulfstein 2000, pp. 623–626). Although direct changes to these external obligations are often made through formal amendments which require ratification, substantive changes can also occur through COP decisions, which are in most cases adopted by consensus without provisions for opt-out, or, if all attempts at consensus fail, some form of majority vote. Through these activities, COPs frequently alter the application and scope of their treaties (Wiersema 2009, pp. 237–245, 286). Some MEAs even contain a catchall provision authorising the COP to fulfil such other functions as may be appropriate under the provisions of the relevant convention. This general authorising provision "tends to provide somewhat open-ended powers related to particular aspects of the treaty" (Wiersema 2009, p. 252). These additional rules emerging inside the institution add up to system-specific bodies of secondary law (Gehring 2008, p. 28). Conventional categories for the sources of International Law cannot capture these activities adequately (Wiersema 2009, p. 232). From a formal standpoint, the legal status of COP decisions is unpleasantly ambiguous. Here, a strict dichotomy between "binding" and "non-binding" International Law breaks down. COP decisions tend to contain mandatory terms and make access to certain benefits like financial aid contingent upon compliance with them. Parties agree to subject themselves to these terms, yet they do not appear to be binding in a formal sense. Thus, the scope of the COP's law-making powers has emerged inside a significant grey zone (Brunnée 2002, p. 32).

The COP model is rare in other fields. Some human rights treaties and some arms control treaties provide for arrangements that, at first glance, might appear to be embryonic forms of the COP model, but they are, in fact, very different. They take place on an ad hoc basis, they have no permanent secretariat, and the role of the meetings of parties is limited to considering possible amendments to the constitutive treaty. They take no real part in supervising the implementation of the treaty concerned. The institutional arrangements of the General Agreement on Tariffs and Trade (GATT) before 1994 come closest to those of the COP model but were largely a historical accident. It was initially assumed that the International Trade Organisation would be created (as an IO) and would provide a more typical institutional machinery for the GATT. But, mainly because of opposition of the United States, the International Trade Organisation never came into being (Churchill and Ulfstein 2000, p. 658).

Although it is unusual in the realm of international institutions more generally, the COP model has a significant advantage, which led to its emergence as the main form of cooperation in international environmental governance, namely its flexibility. In contrast to traditional IOs, the law-making of MEAs does not, as a rule, constitute "independent" secondary legislation but amounts to amending their own constitutive convention and its protocols, annexes, and appendices. This guarantees continuous adaptation to the ever-changing ecological, scientific, and technical context (Sommer 1996, p. 660). Further, political will to address environmental concerns tends to be limited and might take time to consolidate in favour of stricter measures. Structuring MEAs through framework conventions and later added protocols on more specific obligations in the context of regular meetings works in favour of resolving this issue and avoids the cumbersome process of continuously amending a more traditional treaty through ad hoc diplomatic conferences (Churchill and Ulfstein 2000, p. 629).

An Investigation into MEA Authority Expansion

Despite their wide proliferation in the realm of international environmental politics, an issue area that dominates not just the political agenda but our everyday life, the true nature of MEAs as a distinct and specific type of international institution remains little understood. Traditional theories of IR have trouble grasping MEAs as a distinct phenomenon and continue to conceptualise them as mere treaties, functioning as arenas for state interaction. Realists (Waltz 1979; Strange 1982; Mearsheimer 1994) focus on states as the only relevant actors on the global scale, interacting in a constant state of distrust where institutions can never be more than fragile and temporary alliances. Institutionalists (Keohane 1989; Axelrod and Keohane 1985; Abbott and Snidal 1998; Stein 1982; Scharpf 1997) attach much more importance to institutions in general. Still, institutions remain intervening variables with no existence separate from the states they are composed of. Their focus on states as the main centres of authority on the international level precludes them from forming an understanding of MEAs, or international institutions more generally for that matter, as anything more than instruments of state power. The "constructivist turn" in IR theory has included institutions as essential components of the international norm-development process (Wendt 1999; Schmidt 2013; Finnemore and Sikkink 1998; Barnett and Finnemore 1999). However, the focus remains on the rational-legal authority of large-scale bureaucracies and continues to neglect more ad hoc and informal organisations like MEAs.

There is strong reason to believe that these perspectives vastly underestimate MEAs' capabilities to become relevant on the international level in their own right. While MEAs arise from treaties negotiated and adopted by states, the full package of obligations continuously evolves and will expand far beyond the original treaty text (Gehring 2008, p. 18). Although COPs essentially lie at the intergovernmental end of the institutional spectrum, they tend to develop a dynamic of their own which limits state sovereignty. For example, the Bush Administration might have preferred the climate change issue simply to disappear. However, there was no escape from the annual conferences of the UNFCCC. These regular meetings enmesh states in an international process that will ultimately make them act differently as a group than they would on their own. In this way, COPs have impressively shown that they can develop into something much more than a vehicle for transmitting individual state preferences (Bodansky 2010, p. 122).

Many cases can be identified in which MEAs seem to take on a life of their own and expand their capacities beyond what the original treaty prescribed, indeed, beyond what some of the most powerful states would have supported individually. We observe not only that MEAs exert influence at all but that they expand their influence over time, claiming new functions for themselves as they mature. These expansions arise from the inside of MEAs and are adopted by COPs. They are results of a collective process and not secured through formal amendments which require individual national ratification. Some of the most impressive cases will be discussed in later chapters. They include, for example, the establishment of the Interim Multilateral Fund of the Montreal Protocol, which created extensive financial obligations for industrialised country parties which were not at all anticipated when the MEA was set up, or the establishment of the exceptionally strong enforcement mechanism of the Kyoto Protocol's compliance committee which was able to impose sanctions where states failed to achieve their emission targets. Further examples not discussed in detail in this book include the replacement of the prior informed consent procedure of the Basel Convention with a complete ban on exports of hazardous wastes from OECD to non-OECD countries (Kummer 1998, p. 229) or the introduction of various quota systems by the COP of CITES (Wijnstekers 2018, p. 561).

Scholars of International Law have begun to recognise MEAs as a distinct form of cooperation and accept that they are capable of exerting significant influence in their own right (Brunnée 2002; Churchill and Ulfstein 2000; Wiersema 2009), even going so far as to describe them as "virtually identical to international organisations" (Brunnée 2002, p. 16). They agree that the fact that MEAs are established explicitly with a more informal, ad hoc nature should not be a reason to deny them legal personality. Since states can influence the COPs of MEAs only by acting through them, meaning only in a collective rather than an individual capacity, they need to be considered as more than the sum of their members. In essence, they should be regarded as possessing "a will of their own" (Churchill and Ulfstein 2000, p. 658).

Moving to an understanding of MEAs as a type of IO sheds light on the expansive tendencies described above from a theoretical standpoint. IOs are recognised as separate entities with a propensity to dynamically expand on their own, at least by some strands of IR theory. The question of why similar shifts occur in a type of institution that remains widely understood as mere instruments of member states moves to centre stage. Established approaches to the study of IO expansion, however, cannot be directly applied since they tend to focus on the rational-legal authority of large-scale bureaucracies (Hooghe et al. 2019), which MEAs do not possess. Thus, an investigation into MEA-specific driving forces of organisational expansion needs to build on but also go beyond these approaches.

Recent approaches to the study of authority in global governance provide a starting point. A disaggregation of authority in the international landscape away from states as the only focal points and towards a plethora of other actors has been observed (Rosenau 1995, 1992). This new perspective opens up a broader discussion on the concept of authority, how it should be understood and defined, and to whom exactly it can be applied (Lake 2010; Zürn et al. 2012). As Hooghe and Marks (2019, p. 33) have stated, "It is perfectly possible to conceive an authoritative international organisation in which non-state actors play no role at all if the principals collectively make decisions that are binding on individual states".

Different forms of authority in global governance have recently been distinguished. Political authority refers to the transfer of decision-making powers to a governor *in* authority and the acceptance of decisions as collectively binding by the governed. Epistemic authority refers to the acceptance of an actor or organisation as *an* authority due to her expertise (Katsikas 2010, pp. 116–119). Private authority rests between these two categories. Private actors may gain influence based on their expertise and their acceptance by their audience as epistemic authorities (Katsikas 2010, p. 123), but they may also exert political authority, either for a particular community of (usually non-state) actors for which they are authorized to make binding decisions, or based on powers selectively delegated to them by public international institutions (Green 2013, pp. 33–36).

Political authority is a main building block for the conceptualisation and analysis undertaken in this book. Particularly useful for developing a more accurate understanding of MEAs is the concept of relational authority as a form of political authority (Lake 2010). Relational authority is defined as "a social contract in which a governor provides a political order of value to a community in exchange for compliance by the governed with the rules necessary to produce that order" (p. 589). This understanding allows for a much broader application on the international level to include MEAs as separate entities with the ability to exert authority in their own right. Since relational authority is inherently dynamic and subject to constant negotiation and re-negotiation, significant shifts may become possible under suitable circumstances without the need for formal delegation.

Based on this new understanding of MEAs as separate authoritative entities, existing theories of organisational expansion can be adapted to formulate hypotheses on specific driving forces that can make these expansions happen. The groundwork for this line of argumentation is based in the concept of Informal IO Empowerment as an informal version of the IO-focused principal-agent approach. A central mechanism of this approach is conversion, which represents a more gradual and indirect version of the classic agency slack and occurs where rules remain formally the same but are interpreted and implemented in new ways (Heldt and Schmidtke 2017; Mahoney and Thelen 2009). Two hypotheses about the driving forces of these processes will be distinguished based on what is probably the most well-known theory of authority expansion but which is rarely applied outside the European context: Neofunctionalism (Haas 1958; Lindberg 1963; Tranholm-Mikkelsen 1991; Niemann 2006). One is based on the concept of functional spill-over, which assumes a functional need to move a governance function to the collective level to make or keep the MEA fully functional when gaps or ambiguities in the skeletal framework treaty hinder it from proceeding productively. The second is an actor-driven logic loosely based on the concept of cultivated spill-over, which recognises the possibility that, in the absence of functional need, actors inside the MEA might be able to strategically employ the institutional structures to push an expansion which might be in their interest. This includes the use of institutional logics to build favourable arguments as well as the creation of alliances and brokering of package deals.

Six cases of MEA authority expansion without formal delegation will be analysed to investigate these hypotheses and this so far neglected phenomenon more generally in the areas of ozone and climate protection. This includes, in particular, the creation of new funds to provide financial assistance to developing countries, like the creation of the Green Climate Fund (GCF) under the UNFCCC and its Paris Agreement, and the area of compliance, like the creation of the Montreal Protocols compliance management system which introduced a facilitative, preventative approach to the realm of international environmental politics over the use of strict sanctions.

The analysis of these cases provides critical new insights into the inner workings of MEAs and international environmental politics more broadly. Beyond an in-depth understanding of the individual cases and treaties, it will be concluded that functional pressure seems to be the predominant but not the only relevant driving force of authority expansion. Most of the cases analysed here were driven by a functional need to resolve gaps and ambiguities, which needed to be resolved at some point in a continuously evolving regulatory framework. These expansions seem to happen predominantly early on as the skeletal rules of the original treaty are fleshed out through subsequent COP negotiations and slow down as the MEA operates continuously over a longer period of time. Under favourable circumstances, however, an actor-driven logic of authority expansion also seems possible where benefits of a specific expansion are unequally distributed, and those in favour of it can strategically employ the institution's structures to realise it. These insights have significant relevance for our understanding of informal processes of institutional expansion generally and global environmental politics specifically as one of the most pressing issues of our time.

Theoretically, these results advance recent debates on the authority of less formal types of IOs and informal ways of institutional expansion. As we begin to recognise MEAs as separate entities in the global landscape, the spectrum of existing theories on IO expansion, like the principal-agent approach and theories of European Integration, broadens to include more informal ways of similar processes. Further, the findings presented here will provide relevant input into the discussion on authority in global governance more generally by illustrating the inclusion of more informal types of institutional processes and their relevance for state obligations on the international level. This ultimately leads to the question of whether MEAs could even be recognised as actors in their own right, similar to how some have begun to recognise full-fledged IOs as such. While answering this question is beyond the scope of the analysis undertaken here, some have begun to investigate this question based on a perspective that corresponds to the one employed here (Gehring and Urbanski 2023; Gehring and Spielmann 2023).

Empirically, this analysis will provide valuable insights into the inner workings of MEAs and their relevance for state commitments in the broader context of global environmental governance. Many impressive cases, like the creation of the GCF, have been extensively examined with a focus on its institutional design and effectiveness. Still, the specific circumstances of why it was created have, so far, found little attention. Further, some striking cases of authority expansion, like the definition of developing countries under the Montreal Protocol, might only come to light once we specifically look for them.

Practically, states entering this type of commitment should not be surprised to realise that they might have given up more control than expected. However, since MEAs are designed to expand over time to fit the specific needs, knowledge, and compromises that evolve, these expansions should not only be seen as a loss of control but also as an opportunity to adapt to changing circumstances in the most productive way. This is supported by the prevalence of functionally driven cases where the continuous operation of the MEA in question was in jeopardy.

Understanding the specific dynamics at play can improve the efficiency of negotiations and thus positively influence global environmental governance as a whole. When drafting new MEAs or formal amendments to existing ones, states can be more intentional with the language they use to prevent unwanted informal expansions or to prepare for when they occur. Further, recognising the driving force of an informal expansion under discussion allows to view possible paths more clearly and lead to more appropriate and acceptable solutions. Where actor-driven pressure is identified, there will likely be room for different types of deals and compromises. Where functional pressure is at play, the path will be much more narrow and negotiations can focus on purposefully resolving the underlying issue. Recognising a functional need could also help to reduce controversy where informal expansions are unpopular among a number of Parties involved. Recently, COPs have been recognised mostly for their dragging negotiations and unsatisfactory results. The issue of finance seems to be a central point of contention, which has also been identified here as an area where large-scale informal expansions are common. Enhancing clarity, efficiency, and intention in this way thus can help to revive MEAs as central frames of global environmental governance and point of reference under which many smaller public and private initiatives can thrive.

Adjoining Scholarly Debates

Beyond the study of MEAs and authority in global governance, this book draws from and will be of interest to several adjoining debates currently taking place in IR and International Law. The most relevant will be summarised here shortly.

First, the secretariats of MEAs have recently found some scholarly attention. It has been shown that the secretariats of some MEAs have increased their influence on COP-decisions through the orchestration of non-state actors, such as civil society groups, non-profit entities, or the private sector to push for more ambitious responses to environmental problems (Hickmann et al. 2024; Mai and Elsässer 2022; Saerbeck et al. 2020). Still, while this use of informal governance techniques is certainly impressive, it remains undisputed that the role of MEA bureaucracies remains limited in comparison to those of large IOs (Biermann and Sieberhüner 2013; Mai and Elsässer 2022). They have been established as small servicing units with strictly administrative functions (Bodansky 1993, p. 534; Busch 2009, p. 251; Biermann and Siebenhüner 2013, p. 154) and do not seem to influence COP decisions systematically. In the cases analysed here, the main driving forces lie elsewhere. This book complements these insights by providing a larger framework for the study of MEA dynamics beyond secretariat activities and advancing a full picture of factors influencing COP-decision-making in MEAs.

Second is the recent strand of literature investigating institutional adaptation, in particular in MEAs, identifying reasons for why some agreements are adapted frequently while others are not (Laurens et al. 2023; Haftel and Thompson 2018; De Bruyne et al. 2020; Daßler et al. 2019). However, the focus remains on formal instruments like treaty amendments and protocols. Laurens (2023) does include COP-decisions as a form of institutional adaptation, however, her focus lies on how new design features within an MEA framework are generated and introduced and not on the processes leading up to the adoption of these features which are central to this book. Thus, the investigation undertaken here complements these approaches by adding the informal dimension of large-scale expansive COP-decisions to the study of formal MEA adaptation.

Third, a proliferation of different forms of informal governance has been identified in recent years and scholars have begun to analyse their many nuances (Westerwinter et al. 2020; Cooper et al. 2022). This includes "informality *of* institutions, *within* institutions, and *around* institutions" (Westerwinter et al. 2020, p. 5). While informality around institutions lies outside the scope of this book, MEAs fall in-between both other strands. With their founding treaties and extensive institutional machinery, they are more formal than the informal institutions like the G7 or G20, which are the focus of this strand of literature. But, they remain less formal than the large-scale IOs which are the focus of studies of informality within institutions. However, similar processes to the ones uncovered here could take place in both contexts. What most current approaches to informal governance have in common is their main focus on reasons for the emergence of informality, including functional considerations (Abbott and Snidal 2000; Raustiala 2002; Koremenos 2016), power and power asymmetries (Stone 2011, 2013) or the relevance of domestic politics (Kersting and Kilby 2021), regime complexity, and changing causal beliefs (Colgan and Van de Graaf 2015). They focus less on the possible outcomes of these informal processes, which are of central interest to this book. Thus, the analysis undertaken here draws from and provides valuable insights into this debate, but also goes beyond by taking into account institutional expansion as a particularly puzzling outcome of informal processes.

Plan of the Book

The argument of this book is developed in six chapters. It begins with an overview over established theoretical perspectives, then puts forward a conceptualisation of MEAs as dynamic authoritative IOs and, on this basis, develops a theoretical framework for the analysis of informal authority expansions by two driving forces. Six cases of informal authority expansion across two MEAs are analysed to identify their main driving force. Finally, the insights gained are taken together to draw generalised conclusions.

Chapter 1 reviews established perspectives on MEAs in IR and International Law to carve out the significant gap between both sides. Scholars of International Law recognise MEAs as a distinct form of cooperation that functions almost identically to full-fledged IOs. In sharp contrast, the field of IR has trouble grasping MEAs as a distinct phenomenon and continues to conceptualise them as mere treaties, functioning as arenas for state interaction. As a consequence, this perspective significantly underestimates the capabilities of MEAs as described by the law perspective and overlooks how MEAs can expand their authority without formal delegation by their member states.

Chapter 2 develops a conceptualisation of MEAs as organisations from an IR perspective. It begins by reviewing more recent approaches to the study of authority in global governance and the different types of authority that have been distinguished. Focussing on the concept of relational authority, this chapter proceeds to carve out how the COPs of MEAs exercise authority over their individual member states through a dynamic two-dimensional relationship. In this way, MEAs become distinct entities capable of exerting influence on the international landscape in their own right through the performance of different governance functions for their members.

Chapter 3 develops a theoretical framework for the study of informal MEA authority expansion. With relational authority being inherently dynamic, and IOs having a known tendency to expand over time, an investigation into similar processes in MEAs becomes necessary. The basis is provided by the concept of conversion as an informal version of the classic agency slack. To grasp the specific driving forces of informal expansions, two hypotheses are derived by combining relational authority and the neofunctional concepts of functional and cultivated spill-over. This chapter further includes methodological considerations for the application of this framework in subsequent chapters.

Chapters 4 and 5 implement this framework by conducting structured case studies of six instances of authority expansions in two MEAs. Chapter 4 analyses three cases of informal authority expansion under the Vienna Convention on the Protection of the Ozone Layer and the Montreal Protocol on Substances that Deplete the Ozone Layer. This includes the creation of the Interim Multilateral Fund, the decision to classify developing countries on a case-by-case basis, and the creation of the compliance management procedure. All three of these expansions were driven by a functional need for an expansion without ratification. Chapter 5 analyses three cases of informal authority expansion under the UNFCCC with its Kyoto Protocol and Paris Agreement. This includes the creation of the Kyoto Protocol's enforcement mechanism, the creation of the Adaptation Fund Board, and the creation of the GCF. The first and last of these cases are driven by functional need. The creation of the Adaptation Fund Board, however, was driven by cultivated pressure created by developing countries through employing the structures of the Kyoto Protocol in their favour.

The final chapter draws conclusions beyond the individual cases and discusses the broader implications of the results obtained. The insights gained in this book advance our understanding of MEAs and international environmental governance specifically, and discussions on organisational expansion and informal governance more broadly. This chapter further outlines areas of further research to advance and refine the insights gained throughout this book.

References

Abbott, Kenneth W.; Snidal, Duncan (1998): Why States Act through Formal International Organizations. *Journal of Conflict Resolution* 42 (1), pp. 3–32. DOI: 10.1177/0022002798042001001

Abbott, Kenneth W.; Snidal, Duncan (2000): Hard and Soft Law in International Governance. *International Organization* 54 (3), pp. 421–456. DOI: 10.1162/002081800551280

Axelrod, Robert; Keohane, Robert O. (1985): Achieving Cooperation under Anarchy. Strategies and Institutions. *World Politics* 38 (1), pp. 226–254. DOI: 10.2307/2010357

Barnett, Michael N.; Finnemore, Martha (1999): The Politics, Power, and Pathologies of International Organizations. *International Organization* 53 (4), pp. 699–732. DOI: 10.1162/002081899551048

Biermann, Frank; Siebenhüner, Bernd (2013): Problem solving by international bureaucracies: The influence of international secretariats on world politics. In Bob

Reinalda (ed.), *Routledge Handbook of International Organization*. Abingdon and New York: Routledge, pp. 149–161.

Bodansky, Daniel (1993): The United Nations Framework Convention on Climate Change. A Commentary. *Yale Journal of International Law* 18, pp. 451–558.

Bodansky, Daniel (2010): *The Art and Craft of International Environmental Law*. Cambridge, MA.: Harvard University Press.

Brunnée, Jutta (2002): COPing with Consent. Law-Making Under Multilateral Environmental Agreements. *Leiden Journal of International Law* 15 (1), pp. 1–52. DOI: 10.1017/S0922156502000018

Busch, Per-Olof (2009): The climate secretariat. Making a living in a straitjacket. In Frank Biermann, Bernd Siebenhüner (eds.): *Managers of Global Change. The Influence of International Environmental Bureaucracies*. Cambridge, MA: MIT Press, pp. 245–264.

Churchill, Robin R.; Ulfstein, Geir (2000): Autonomous Institutional Arrangements in Multilateral Environmental Agreements. A Little-Noticed Phenomenon in International Law. *The American Journal of International Law* 94 (4), p. 623. DOI: 10.2307/2589775

Colgan, Jeff D.; Van de Graaf, Thijs (2015): Mechanisms of Informal Governance: Evidence from the IEA. *Journal of International Relations and Development* 18, pp. 455–481, https://doi.org/10.1057/jird.201

Cooper, Andrew F.; Parlar Dal, Emel; Cannon, Brendon (2022): The Cascading Dynamics of Informal Institutions: Organizational Processes and Governance Implications. *International Politics*, pp. 1–22, https://doi.org/10.1057/s41311-022-00399-4

Daßler, Benjamin; Kruck, Andreas; Zangl, Bernhard (2019): Interactions Between Hard and Soft Power: The Institutional Adaptation of International Intellectual Property Protection to Global Power Shifts. *European Journal of International Relations*, 25 (2), pp. 588–612. https://doi.org/10.1177/1354066118768871

de Bruyne, Charlotte; Fischenhendler, Itay; Haftel, Yoram (2020): Design and Change in Transboundary Freshwater Agreements. *Climatic Change* 162 (2), pp. 321–341.

Finnemore, Martha; Sikkink, Kathryn (1998): International Norm Dynamics and Political Change. *International Organization* 52 (4), pp. 887–917. DOI: 10.1162/002081898550789

Gehring, Thomas (2008): Treaty-Making and Treaty Evolution. In Daniel Bodansky, Jutta Brunnée, Ellen Hey (eds.): *The Oxford Handbook of International Environmental Law*. Oxford: Oxford University Press, pp. 467–497.

Gehring, Thomas; Spielmann, Linda (2023): The Treaty Management Organization Established Under the UNFCCC and the Paris Agreement: An International Actor in Its Own Right? *International Environmental Agreements* 23, pp. 235–252. DOI: 10.1007/s10784-023-09611-z

Gehring, Thomas; Urbanski, Kevin (2023): Member-Dominated International Organizations as Actors: A Bottom-Up Theory of Corporate Agency. *International Theory* 15 (1), pp. 129–153. DOI: 10.1017/S1752971922000069

Green, Jessica F. (2013): *Rethinking Private Authority. Agents and Entrepreneurs in Global Environmental Governance*. Princeton, NJ: Princeton University Press.

Haas, Ernst B. (1958): *The Uniting of Europe. Political, Social, and Economical Forces: 1950–1957*. Notre Dame, IN: University of Notre Dame Press.

Haftel, Yoram Z.; Thompson, Alexander (2018): When Do States Renegotiate Investment Agreements? The Impact of Arbitration. *The Review of International Organizations* 13 (1), pp. 25–48.

Heldt, Eugénia; Schmidtke, Henning (2017): Measuring the Empowerment of International Organizations. The Evolution of Financial and Staff Capabilities. *Global Policy* 8 (5), pp. 51–61. DOI: 10.1111/1758-5899.12449

Hickmann Thomas; Widerberg Oscar; Lederer, Markus ; Pattberg, Philipp (2024): The Evolution of International Environmental Bureaucracies: How the Climate Secretariat Is Loosening Its Straitjacket. In: Jörgens H, Kolleck N, Well M, (eds.) *International Public Administrations in Environmental Governance: The Role of Autonomy, Agency, and the Quest for Attention*. Cambridge: Cambridge University Press, pp. 57–72.

Hooghe, Liesbet; Tobias Lenz; Marks, Gary (2019): *A Theory of International Organization*, Oxford: Oxford University Press. https://doi.org/10.1093/oso/9780198766988.001.0001

Katsikas, Dimitrios (2010): Non-State Authority and Global Governance. *Review of International Studies* 36 (S1), pp. 113–135. DOI: 10.1017/S0260210510000793

Keohane, Robert O. (1989): *International Institutions and State Power. Essays in International Relations Theory*. Boulder, CO.: Westview Press.

Kersting, Erasmus; Kilby, Christopher (2021): Do Domestic Politics Shape U.S. Influence in the World Bank?. *Review of International Organisations* 16, pp. 29–58. DOI: https://doi.org/10.1007/s11558-018-9321-8

Koremenos, Barbara (2016). *The Continent of International Law. Explaining Agreement Design*. Cambridge: Cambridge University Press.

Kummer, Katharina (1998): The Basel Convention: Ten Years On. *Review of European Community & International Environmental Law* 7 (3), pp. 227–236. DOI: 10.1111/1467-9388.00154

Lake, David A. (2010): Rightful Rules. Authority, Order, and the Foundations of Global Governance. *International Studies Quarterly* 54 (3), pp. 587–613. DOI: 10.1111/j.1468-2478.2010.00601.x

Laurens, Noémie (2023): Institutional Adaptation in Slow Motion: Zooming In on Desertification Governance. *Global Environmental Politics*, 23 (2), pp. 31–53. DOI: https://doi.org/10.1162/glep_a_00705

Laurens, Noémie; Hollway, James; Morin, Jean-Frédéric (2023): Checking for Updates: Ratification, Design, and Institutional Adaptation. *International Studies Quarterly*, 67 (3), pp. 1–13. DOI: https://doi.org/10.1093/isq/sqad049

Lindberg, Leon N. (1963): *The Political Dynamics of European Economic Integration*. Stanford, CA: Stanford University Press.

Mahoney, James; Thelen, Kathleen (2009): *Explaining Institutional Change*. Cambridge: Cambridge University Press.

Mai, Laura; Elsässer, Joshua (2022): Orchestrating Global Climate Governance Through Data: The UNFCCC Secretariat and the Global Climate Action Platform. *Global Environmental Politics*, 22 (4), pp. 151–172. DOI: https://doi.org/10.1162/glep_a_00667

Mearsheimer, John J. (1994): The False Promise of International Institutions. *International Security* 19 (3), pp. 5–49. DOI: 10.2307/2539078

Niemann, Arne (2006): *Explaining Decisions in the European Union*. Cambridge: Cambridge University Press.

Oberthür, Sebastian; Gehring, Thomas (2004): Reforming International Environmental Governance. An Institutionalist Critique of the Proposal for a World Environment Organisation. *International Environmental Agreements* 4 (4), pp. 359–381. DOI: 10.1007/s10784-004-3095-6

Rajamani, Lavanya (2012): The Changing Fortunes of Differential Treatment in the Evolution of International Environmental Law. *International Affairs* 88 (3), pp. 605–623. DOI: 10.1111/j.1468-2346.2012.01091.x

Raustiala, K. (2002). The Architecture of International Cooperation: Transgovernmental Networks and the Future of International Law. *Virginia Journal of International Law*, 43 (1), pp. 1–92.

Rosenau, James N. (1992): Governance, order, and change in world politics. In Ernst Otto Czempiel, James N. Rosenau (Eds.): *Governance Without Government. Order and Change in World Politics*. Cambridge: Cambridge University Press (vol 20), pp. 1–29.

Rosenau, James N. (1995): Governance in the Twenty-first Century. *Global Governance* 1 (1), pp. 13–43. DOI: 10.1163/19426720-001-01-90000004

Saerbeck, Barbara; Well, Mareike; Jörgens, Helge; Goritz, Alexandra; Kolleck, Nina (2020): Brokering Climate Action: The UNFCCC Secretariat Between Parties and Nonparty Stakeholders. *Global Environmental Politics*, 20 (2), pp. 105–127. DOI: https://doi.org/10.1162/glep_a_00556

Scharpf, Fritz W. (1997): *Games Real Actors Play. Actor-Centered Institutionalism in Policy Research*. Boulder, CO.: Westview Press.

Schmidt, Brian C. (2013): On the History and historiography of international relations. In Walter Carlsnaes, Thomas Risse, Beth A. Simmons (eds.): *Handbook of International Relations*. Los Angeles, CA.: Sage, pp. 3–28.

Sommer, Julia (1996): Environmental Law-Making by International Organisations. *Zeitschrift für Ausländisches Öffentliches Recht und Völkerrecht* 56, pp. 628–667.

Stein, Arthur (1982): Coordination and Collaboration: Regimes in an Anarchic World. *International Organization* 36 (2), pp. 299–324.

Stone, Randall W. (2011): *Controlling Institutions: International Organizations and the Global Economy*. Cambridge: Cambridge University Press.

Stone, Randall W. (2013): Informal Governance in International Organization: Introduction to the Special Issue. *Review of International Organizations*, 8 (2), pp. 121–136.

Strange, Susan (1982): Cave! Hic Dragones: A Critique of Regime Analysis. *International Organization* 36 (2), pp. 479–496. DOI: 10.1017/S0020818300019020

Tranholm-Mikkelsen, Jeppe (1991): Neo-Functionalism. Obstinate or Obsolete? A Reappraisal in the Light of the New Dynamism of the EC. *Millennium* 20 (1), pp. 1–22. DOI: 10.1177/03058298910200010201

Waltz, Kenneth N. (1979): *Theory of International Politics*. Reading, MA: Addison-Wesley.

Wendt, Alexander (1999): *Social Theory of International Politics*: Cambridge: Cambridge University Press.

Westerwinter, Oliver; Abbott, Kenneth W.; Biersteker, Thomas (2020): Informal Governance in World Politics. *Review of International Organisations*, 16, pp. 1–27. DOI: https://doi.org/10.1007/s11558-020-09382-1

Wiersema, Annecoos (2009): The New International Law-Makers? Conferences of the Parties to Multilateral Environmental Agreements. *Michigan Journal of International Law* 31 (1), pp. 231–287.

Wijnstekers, Willem (2018): *The Evolution of CITES: A Reference to the Convention on International Trade in Endangered Species of Wild Fauna and Flora*: 11th edition. Budakeszi, Hungary: International Council for Game and Wildlife Conservation (CIC).

Zürn, Michael; Binder, Martin; Ecker-Ehrhardt, Matthias (2012): International Authority and Its Politicization. *International Theory* 4 (1), pp. 69–106. DOI: 10.1017/S1752971912000012

1 Towards a New Understanding of MEAs

Multilateral Environmental Agreements (MEAs) constitute the predominant institutional form of public global governance in a policy field that lacks a broadly responsible International Organisation. They are structured in a peculiar hybrid form in-between treaties and organisations and operate through a horizontal structure in which member states delegate tasks to a collective of themselves in the form of a Conference of the Parties (COP). Under strict state control in a formal sense, they comprise impressive institutional machinery for secondary decision-making along all stages of the regulatory process. These structures amount to much more than what a treaty would normally possess but not to the same large-scale bureaucracies of traditional International Organisations. The true nature of MEAs, thus, is hard to grasp and classify. Scholars of International Law have begun to recognise them as powerful players in their own right and accept them to closely resemble organisations. The field of International Relations (IR), however, is only slowly beginning to catch up. Traditional approaches in the field understand MEAs as arenas for state interaction with no capabilities of their own. Moving beyond these state-centric theories, the global governance approach opens up the discussion to allow for a more nuanced analysis. It observes a disaggregation of authority in the international landscape, away from states as the only focal points and towards a plethora of other entities who can become essential players in their own right. However, those following this approach remain hesitant to come to the same conclusion on MEAs as those in the field of law. This chapter summarises the law perspective and brings it face to face with the predominant views of IR theory as well as the emerging debate on authority in global governance.

International Law and MEAs as International Organisations

Scholars of Internal Law have extensively discussed the peculiar nature and capabilities of MEAs (Brunnée 2002; Churchill and Ulfstein 2000; Wiersema 2009). From their perspective, MEAs should be understood as a distinct form of organisation which can actively engage in law-making on the international

DOI: 10.4324/9781003597681-2

level. They tend to agree that MEAs are "virtually identical to international organisations" (Brunnée 2002, p. 16) and that despite their more ad hoc nature, they establish:

> independent intergovernmental bodies (COPs) with the potential to develop powers over states that may far exceed those of more formally established institutions.
>
> (Werksman 2014, p. 55)

While this opinion stands in sharp contrast to established perspectives of IR theory, it is supported by an opinion of the UN Office of Legal Affairs where it was stated on November 4, 1993 that the Climate Change Convention established "an international entity/organisation with its own separate legal personality, statement of principles, organs and a supportive structure in the form of a Secretariat"; and, in an opinion of December 18, 1995, it added that the bodies established by this convention "have certain distinctive elements attributable to international organisations" (UN Doc. FCCC/SBI/1996/7, para. 11 (2) (1996)).

Indeed, there is no legal or generally accepted definition of what constitutes an "International Organisation". While scholars of IR tend to focus on large-scale bureaucracies as central characteristic, the International Law Commission agreed upon the following:

> the term 'international organisation' refers to an organisation established by a treaty or other instrument governed by international law and possessing its own international legal personality.
>
> (UN 1966, p. 190)

Largely drawing on this statement, Schermers and Blokker (2011) define (public) International Organisations as "forms of cooperation founded on an international agreement, creating at least one organ with a will its own, established under international law" (p. 37).

While the condition of being founded on an international agreement or treaty can be clearly accepted in relation to MEAs, the second part is more troubling. International legal personality used to be a hallmark of sovereign state power, but it has now been recognised that there is no necessary connection between both. Accordingly, alongside an increasing role for International Organisations, a general acceptance of IOs as separate legal entities has gradually emerged (Sands et al. 2009, p. 473). Legal personality means that an organisation has the capacity to bear rights and obligations, which allows it to function as an independent unit and to participate meaningfully in international legal life. In essence, it can exist in the legal sphere. It follows that if an entity is accepted to be an International Organisation, it is also accepted to be much more than an ad hoc grouping of states but as a separate body to which decision-making power has been delegated. Such a body, therefore, is much more than merely a sum of its members (Schermers and Blokker 2011, p. 44).

Scholars of International Law have extensively discussed whether MEAs should be subsumed under this condition. They conclude that the COP of an MEA, as well as its subsidiary bodies, should indeed be understood as possessing legal personality or a "will of their own" since states may influence their work only by acting through them (Churchill and Ulfstein 2000, p. 633). Few MEAs to date have explicitly designated their institutions as international legal persons. Still, such an explicit designation is not necessary under International Law and can also be implied through an entity's nature and functions (Scott 2013). Thus, even though MEAs have been consciously established in a more informal way, this should not be a reason to deny them legal personality. The reason for their more ad hoc design can be seen more in the need for "institutional economy" than a desire to establish less-effective institutions. It follows that "State parties should therefore not be surprised to learn that they have allocated more powers to MEA institutions than they have expressly conferred on them" (Churchill and Ulfstein 2000, p. 649).

It follows that, while due account should be given to their specific nature, an accurate assessment of MEA powers needs to go beyond the consent-based structures of treaty law as such (Gehring 1990; Brunnée 2002; Gehring 2008) and should be supplemented by the much broader principles of international institutional law (Churchill and Ulfstein 2000; Sands et al. 2009; Sands 2003). Since COPs represent hybrids between issue-specific diplomatic conferences and permanent plenary bodies of International Organisations, MEAs exist at the interface of both (Brunnée 2002, p. 16). MEAs are, themselves, treaties, but COP decisions are not. The functions that COPs perform are similar to those of IOs. As a result, the law of treaties does not directly apply, and international institutional law is needed to fill these gaps (Churchill and Ulfstein 2000, p. 633). This clearly shows that an understanding of MEAs as mere treaties inevitably falls short of grasping their true nature to its full extent.

A central consequence of applying institutional law is the possible application of the doctrine of implied powers. The doctrine of implied powers emphasises the object and purpose of IOs and provides a legal basis for decision-making that is not explicitly provided for in the original agreement (Ulfstein 2008, p. 881). Its foundations lie within the Reparations for injuries suffered in the service of the United Nations Advisory Opinion (Reparations Case). Here, the International Court of Justice was confronted with the question of whether the UN possessed the capacity to bring an international claim in respect of damage caused to the UN and to the victims or persons entitled through them. No such competence has been explicitly given to the UN in the Charter (Schermers and Blokker 2011, p. 182). In its advisory opinion of April 11, 1949, the Court held:

> Under international law, the Organisation must be deemed to have those powers which, though not expressly provided in the Charter, are conferred

> upon it by necessary implication and being essential to the performance of its duties.

The extent of implied powers can, therefore, be defined in a functional way by asking if the power in question is necessary or essential for the organisation to fulfil its purpose (Naturally, opinions on this subject might differ) (Schermers and Blokker 2011, p. 182).

> In international organisations, the doctrine of implied powers means that the organisation is deemed to have certain powers which are additional to those expressly stipulated in the constituent instrument. These additional powers are necessary or essential for the fulfilment of the tasks or purposes of the organisation, or for the performance of its functions, or for the exercise of the powers explicitly granted.
>
> (Skubiszewski 1989, p. 856)

If MEAs are recognised as similar to IOs, they may also be recognised as enjoying implied powers in relation to the international level as well as in relation to matters concerning substantive obligations of their member state parties. Indeed, as will be illustrated in later chapters, COPs have taken various law-making decisions that were intended to be and seem to be regarded as legally binding without express provision in the original treaty. These decisions could be considered as being based on implied powers. This is, however, a somewhat uncertain foundation, and a party unhappy with such a decision could always argue that the COP was acting ultra vires. This problem does not loom as large for those MEAs that contain a "catch-all provision" authorising COPs to "consider any additional action that may be required", "fulfil such other functions as may be appropriate under the provisions of the Convention", or "exercise such other functions as may be appropriate under the provisions of the objective of the Convention", and thus explicitly permitting any action to achieve the purpose of the agreement (Churchill and Ulfstein 2000, pp. 633–634).

In this context, the legal nature of COP decisions is unpleasantly ambiguous. A strict dichotomy between "binding" and "non-binding" International Law breaks down in the context of MEAs. COP decisions tend to contain terms that make them mandatory ("shall") and make access to certain benefits like financial aid contingent upon compliance with some of these mandatory terms. Parties agree to subject themselves to their terms, yet they do not appear to be binding in a formal sense. Thus, the scope of the COP's law-making powers has emerged inside a significant grey zone (Brunnée 2002, p. 32).

Many have tried to resolve this issue by characterising COP decisions as "soft law". However, the concept of soft law has been widely criticised in the field of International Law for its lack of coherent theoretical underpinning and support in state practice. More importantly, as Wiersema (2009, p. 264) argues, the depiction of COP activity as soft law obscures its true nature since decision-making is so tightly bound up with the obligations of the MEAs constitutional treaty that it does not fit in with these categories created for independently emerging international norms. As such, the question of the true legal status of COP decisions evades a clear answer.

Moving away from a strictly formal perspective on International Law, Brunnée (2002) proposes an interactional perspective on International Law-making more generally and MEA law-making specifically. Her perspective is informed by constructivist IR theory and conceives of International Law as arising "from a mutually generative process" (p. 34). States, through their interaction, influence the scope and content of international norms and institutions, which in turn furnish the context within which further state interaction takes place (pp. 33–35). In this view, law-making occurs along a continuum of (formally) binding and non-binding outputs, which includes the "grey zone" of COP decisions. Thus, MEAs can engage in legislative activity whether or not their decisions are binding in a formal sense. In this interactional perspective, legislation is never a unidirectional imposition of authority, but it is, by definition, contingent upon mutually generative activity and reciprocity of expectations between lawgivers and subjects.

> Legislation, then, is not law because it was produced by a "legislature" in the conventional sense but because it was generated through a successful interactive process. Therefore, when we think of COPs as legislatures, we should think of them as collectives that are engaged in law-making in this richer sense, rather than in the purely formal sense.
>
> (p. 38)

Wiersema (2009) goes even further by not only recognising COPs as a distinct but significant source of international legal obligation but also recognising the consequences of independent MEA activity. In the absence of opt-out provisions, the consensus-based law-making activities of COPs diminish the role of individual state consent. In the context of the high level of adaptability of an MEAs skeletal framework, thus, COPs are empowered with significant responsibility of their own, which might lead to shifts of an MEAs focus where control shifts away from individual states towards the collective. In this context, they become increasingly self-referential and disregard the broader international legal context (pp. 271–276).

These perspectives resonate with the conceptualisation of MEAs based on more recent approaches in the study of IR that will be carved out in later chapters. They lay important groundwork for an understanding of MEAs as more than empty shells of state interaction and shift our focus towards a conceptualisation of MEAs as active players in International Law-making.

Inter-National Relations, Global Governance, and MEAs as Arenas for State Interaction

In sharp contrast to the law perspective, the field of IR continues to understand MEAs as mere treaties, set up to structure state interaction. Despite their widespread proliferation in the realm of international environmental politics, an issue area that dominates not just the political agenda but our everyday life, the true nature of MEAs as a distinct and specific type of international institution remains little understood in this field.

Traditional theories of IR, like realism (Waltz 1979; Strange 1982; Mearsheimer 1994) and neoliberal institutionalism (Keohane 1989; Axelrod and Keohane 1985; Abbott and Snidal 1998; Stein 1982; Scharpf 1997) have trouble grasping MEAs as a distinct phenomenon and continue to conceptualise them as mere treaties, functioning as arenas for state interaction. The focus on states as the main centres of authority on the international level precludes these theories from forming an understanding of MEAs, or international institutions more generally, as anything more than instruments of state power. Thus, MEAs remain intervening variables with no existence separate from the states they are composed of.

To conceptualise MEAs, the neoliberal notion of international regimes is widely applied. Stephen D. Krasner (1982) famously defined regimes as "sets of implicit or explicit principles, norms, rules, and decision-making procedures around which actors' expectations converge in a given area of international relations" (p. 186). This definition not only includes traditional formal IOs but more broadly applies to most types of institutionalised cooperation on the international level. Regimes are understood as "intervening variables standing between basic causal factors on the one hand and outcomes and behaviour on the other" (p. 185). They are structures mediating between the state's pursuit of self-interest and political outcomes by changing the structure of opportunities and constraints. Still, while regimes are relevant and essential components of the international system, they have no purpose independent of the states that comprise them (Barnett and Finnemore 1999; Keohane 1984; Baldwin 1993). The institutional machinery that might come with a regime is simply seen as one feature of the wider institutional setting but is not looked at more closely (Bauer 2006, p. 26). Thus, while this perspective recognises MEAs and institutions more generally as more than mere epiphenomena of power relations, they still remain passive instruments of state action.

More recent approaches in the tradition of institutionalism have opened the door for an understanding of institutions as something more. The principal-agent approach originated in economic analyses of corporations and focuses on the role of institutional frames and contracts for the behaviour of individual and collective actors while emphasising the dominance of self-interested and opportunistic motivations of all actors (Bauer et al. 2009, pp. 26–27; Williamson 1985). Principal-agent theory models strategic interaction between a principal who delegates a specific set of tasks to an agent who receives a conditional grant of authority to act on the principal's behalf to fulfil these tasks. Central to principal-agent theory is the assumption that agents will act opportunistically and pursue their own individual interests subject to constraints imposed by the principals (Kiewiet and McCubbins 1992, p. 5). Bringing together economic and institutional theory, the principal-agent approach conceptualises the relationship between IOs and their member states as similar to that between corporate management and shareholders. The principals, meaning the shareholders of a company or the member states of an IO, delegate a specified set of tasks to the management of a company or the bureaucracy of an IO. The management, or the IO bureaucracy, then fulfils this set of tasks with the necessary degree of autonomy in everyday decision-making. In this constellation, conflicts of interest and asymmetric information might emerge (Bauer et al. 2009, p. 26). Even after control mechanisms are established, a range of potential independent actions will remain available for agents in fulfilling the delegated task, thus leaving room for self-interested decisions and inevitably causing some agency loss for the principals. This loss, termed agency slack, describes a situation in which the agents use their autonomy to favour their own interest over that of the principals – thus developing a life of their own (Hawkins et al. 2006, p. 8). What bureaucracies are interested in is similar to what managers might enjoy maximising: Power, prestige, and amenities. Therefore, they aim to "maximise their budget, staff and independence" (Vaubel 1996, p. 195). Principal-agent theory thus maintains that international bureaucracies are able to act autonomously of their principals and, therefore, need to be conceptualised as actors in their own right (Bauer et al. 2009, p. 26). While this is a significant step towards recognising international institutions as more than mere arenas for state interaction, the principal-agent approach relies heavily on the specifics of full-fledged IOs and thus is hard to translate into the context of MEAs which lack precisely the large-scale bureaucracies and hierarchical structure the principal-agent approach identifies as a source of IO agency.

The "constructivist turn" in IR theory has included institutions as essential components of the international norm-development process (Wendt 1999; Schmidt 2013; Finnemore and Sikkink 1998; Barnett and Finnemore 1999). However, the focus remains on the rational-legal authority of large-scale bureaucracies and continues to neglect more ad hoc and informal organisations like MEAs. Fuelled by this development, some institutionalists have begun to draw on approaches originating from the field of sociology. Sociological institutionalism aims to look beyond the "limits of rationality" and seeks to integrate symbols, cognitive scripts, and moral templates into the framework of institutional characteristics (Bauer 2006, p. 27). This approach questions the instrumentalist view of international institutions. It replaces the rationalist interest-based "logic of consequentiality" as the main driving force of behaviour with the normative "logic of appropriateness" (March and Olsen 1998, p. 949). Sociological institutionalism differs from more radical strands of constructivism in that it acknowledges that rationality and norms are intimately connected. Thus, they seek to combine and not replace established explanatory categories of interest and power with the analysis of norms and ideas and integrate an idealist ontology in a positivist epistemology (Wendt 1999; Schmidt 2013; Barnett and Finnemore 1999; Finnemore and Sikkink 1998). Once the focus shifts to normative structures as the causal factors in world politics, all those who create, shape, and maintain such normative structures must be considered relevant participants in the international policy-making process. This also includes all those who rely not on material but rather on ideational resources such as legitimacy, credibility, knowledge, and information. If state interests can be changed through normative structures, then non-state actors who focus on normative development gain relevance and thus move into the centre of sociological institutionalist analysis. IOs – or, more precisely, their bureaucracies – are considered to take part in this norm development process and are thus also recognised as actors in their own right from the constructivist perspective (Barnett and Finnemore 1999, 2004). These capabilities are attributed to their "bureaucratic personality" – as opposed to the legal personality, which is commonly generated from a document of International Law. Bureaucratic personality is constituted by the collective of civil servants who first and foremost serve the organisation's objectives and not so much the partial interests of the member states they are individually affiliated with by nationality (Bauer 2006, p. 29). MEAs could be understood as having normative relevance on the international level. However, their institutional machinery does not compare to the large-scale bureaucracies of traditional IOs. Thus, while further advancing our understanding of IOs as separate entities on the international level, these approaches once again do not provide a basis for an understanding of MEAs as anything more than mere treaties concluded by states.

While the very idea of "inter-national" relations is by definition primarily interested in "politics among nations" (Morgenthau 1948), the global governance approach attaches equal importance to non-governmental organisations (NGOs), transnational corporations, and scientific actors. The term "governance", in this context, can be understood as "the exercise of authority by an actor over some limited community", which can be performed by governments as well as many other types of actors (Lake 2010, p. 590). Thus, where authority was once only attributed to the rational self-interest of states alone, the global governance perspective acknowledges that it is spreading to new realms of world politics (Dingwerth and Pattberg 2006; Czempiel and Rosenau 1992). Still, scholars of global governance have been reluctant to highlight the concept of authority for fear of being dismissed by others committed to the centrality of anarchy at the international level. "If governance is the exercise of authority, and international politics is anarchic or devoid of authority, then there can be no such thing as global governance" (Lake 2010, p. 591). However, global governance should be understood, in its essence, as a set of authority relationships not just among states but among a number of different types of actors, including NGOs, religious orders, private transnational firms, and International Organisations. Once the centrality of authority for international relations is recognised, the door opens analytically to a much wider scope (Lake 2010, p. 592).

James N. Rosenau (1995) famously argued that a new form of anarchy has evolved that includes not only the absence of a highest authority but also encompasses an extensive disaggregation of authority on the international level away from states and towards other types of actors that have not been taken into account before. This development was at least partly triggered by the end of the Cold War which lifted the constraints of a bipolar world order of superpower competition. Rosenau states that a "tendency can be identified in which major shifts in the location of authority and the site of control mechanisms are underway" (p. 18) and that a consequence of this process is the:

> lessening (of) the capacities for governance located at the level of sovereign states and national societies. Much governance will doubtless continue to be sustained by states and their governments (…) but the effectiveness of their policies is likely to be undermined.
>
> (p. 19)

He later describes a disaggregation of hegemonic state power on the international level into multiple "spheres of authority" (2007, p. 88), which can consist of formally recognised states but also of informal networks, broad advocacy networks, as well as narrow, specialised interest organisations, corporations, and NGOs.

While Rosenau's approach expects MEAs to be undermined along with the states that created them, the literature on regime complexity finds that they continue to provide important focal points for the interactions of different levels of global environmental governance. Green (2013a) finds a high level of recognition of public authority by private regulators. She describes the Kyoto Protocol as a "coral reef" (p. 2) in the area of climate governance that attracts private rule-makers whose activities form part of the regime complex to contribute to an orderly expansion of the regime complex. Thus, she concludes that "its vestiges will likely remain, irrespective of the outcome of the inter-governmental process, perpetuated through private authority" (p. 2). Similarly, Betsill et al. (2015) propose to use notions of complex systems to reconceptualise the UNFCCC as a coordinating node in a diverse landscape of initiatives. They suspect different types of linkages between the UNFCCC and other types of governance arrangements to organise divisions of labour and to catalyse action. Hickmann (2017) even states that "the impact of transnational initiatives relies on the continuous evolution of the norms and rules set out in international climate agreements" (p. 445). He observes a "reconfiguration of authority" (p. 432) in which transnational initiatives and state-based forms of governance reinforce each other. In line with Betsill et al.

(2015, p. 3), he finds that "the multilateral treaty process has comparative advantages in delivering some governance functions that are essential to the functioning of the overall system".

Thus, while the global governance perspective significantly broadens the group of relevant actors on the international level, the nature of MEAs in this context remains contested and unclear. They seem to remain arenas of interaction, now structuring not only state behaviour but also the interaction of newly emerging centres of authority. However, taking into account their extensive machinery and bold actions, as well as the striking position of scholars of International Law on their legal standing, there is reason to believe that they are much more than that. With governance being defined as the exercise of authority, a deeper dive into the concept of authority and its varying facets and definitions is warranted to arrive at a clear stance on the true nature of MEAs from a global governance perspective. While more traditional understandings of authority are tied to institutional form and thus have limited capacities, a newer understanding of the concept is emerging, which takes a step beyond these boundaries and can be advanced to encompass a much larger number of institutions on the international level, including MEAs. The following chapter retraces the emergence of this debate and shows how the concept of authority applies to MEAs specifically.

References

Abbott, Kenneth W.; Green, Jessica F.; Keohane, Robert O. (2016): Organizational Ecology and Institutional Change in Global Governance. *International Organization* 70 (2), pp. 247–277. DOI: 10.1017/S0020818315000338

Abbott, Kenneth W.; Snidal, Duncan (1998): Why States Act through Formal International Organizations. *Journal of Conflict Resolution* 42 (1), pp. 3–32. DOI: 10.1177/0022002798042001001

Axelrod, Robert; Keohane, Robert O. (1985): Achieving Cooperation under Anarchy. Strategies and Institutions. *World Politics* 38 (1), pp. 226–254. DOI: 10.2307/2010357

Baldwin, David A. (1993): *Neorealism and Neoliberalism. The Contemporary Debate*. New York, NY: Columbia University Press.

Barnett, Michael N.; Finnemore, Martha (1999): The Politics, Power, and Pathologies of International Organizations. *International Organization* 53 (4), pp. 699–732. DOI: 10.1162/002081899551048

Barnett, Michael N.; Finnemore, Martha (2004): Rules for the World: International Organizations in Global Politics. Ithaca, NY: Cornell University Press.

Bauer, Steffen (2006): Does Bureaucracy Really Matter? The Authority of Intergovernmental Treaty Secretariats in Global Environmental Politics. *Global Environmental Politics* 6 (1), pp. 23–49. DOI: 10.1162/glep.2006.6.1.23

Bauer, Steffen; Biermann, Frank; Dingwerth, Klaus; Siebenhüner, Bernd (2009): Understanding International Bureaucracies. Taking Stock. In Frank Biermann, Bernd Siebenhüner (Eds.): *Managers of Global Change. The Influence of International Environmental Bureaucracies*. Cambridge, MA: MIT Press, pp. 15–36.

Betsill, Michele; Dubash, Navroz K.; Paterson, Matthew; van Asselt, Harro; Vihma, Antto; Winkler, Harald (2015): Building Productive Links between the UNFCCC and the Broader Global Climate Governance Landscape. *Global Environmental Politics* 15 (2), pp. 1–10. DOI: 10.1162/GLEP_a_00294

Brunnée, Jutta (2002): COPing with Consent. Law-Making Under Multilateral Environmental Agreements. *Leiden Journal of International Law* 15 (1), pp. 1–52. DOI: 10.1017/S0922156502000018

Churchill, Robin R.; Ulfstein, Geir (2000): Autonomous Institutional Arrangements in Multilateral Environmental Agreements. A Little-Noticed Phenomenon in International Law. *American Journal of International Law* 94 (4), p. 623. DOI: 10.2307/2589775

Cooper, Scott; Hawkins, Darren; Jacoby, Wade; Nielson, Daniel (2008): Yielding Sovereignty to International Institutions. Bringing System Structure Back In. *International Studies Review* 10 (3), pp. 501–524. DOI: 10.1111/j.1468-2486.2008.00802.x

Czempiel, Ernst Otto; Rosenau, James N. (Eds.) (1992): *Governance without Government. Order and Change in World Politics*. Cambridge: Cambridge University Press (20).

Dingwerth, Klaus; Pattberg, Philipp (2006): Global Governance as a Perspective on World Politics. *Global Governance* 12 (2), pp. 185–204. DOI: 10.1163/19426720-01202006

Finnemore, Martha; Sikkink, Kathryn (1998): International Norm Dynamics and Political Change. *International Organization* 52 (4), pp. 887–917. DOI: 10.1162/002081898550789

Gehring, Thomas (1990): International Environmental Regimes. Dynamic Sectoral Legal Systems. *Yearbook of International Environmental Law* 1 (1), pp. 35–56. DOI: 10.1093/yiel/1.1.35

Gehring, Thomas (2008): Treaty-Making and Treaty Evolution. In Daniel Bodansky, Jutta Brunnée, Ellen Hey (Eds.): *The Oxford Handbook of International Environmental Law*. Oxford: Oxford University Press.

Green, Jessica F. (2013a): Order out of Chaos. Public and Private Rules for Managing Carbon. *Global Environmental Politics* 13 (2), pp. 1–25. DOI: 10.1162/GLEP_a_00164

Hall, Rodney Bruce; Biersteker, Thomas J. (2002): The Emergence of Private Authority in the International System. In Rodney Bruce Hall, Thomas J. Biersteker (Eds.): *The Emergence of Private Authority in Global Governance*. Cambridge: Cambridge University Press (85), pp. 3–22.

Hawkins, Darren G.; Lake, David A.; Nielson, Daniel L.; Tierney, Michael J. (2006): *Delegation and Agency in International Organizations*. Cambridge: Cambridge University Press.

Hickmann, Thomas (2017): The Reconfiguration of Authority in Global Climate Governance. *International Studies Review* 19 (3), pp. 430–451. DOI: 10.1093/isr/vix037

Keohane, Robert O. (1984): *After Hegemony. Cooperation and Discord in the World Political Economy*. Princeton, NJ: Princeton University Press.

Keohane, Robert O. (1989): *International Institutions and State Power. Essays in International Relations Theory*. Boulder, CO: Westview Press.

Kiewiet, D. Roderick.; McCubbins, Mathew D. (1992): The Logic of Delegation: Congressional Parties and the Appropriations Process, by D. Roderick Kiewiet and Mathew D. McCubbins. *Political Science Quarterly* 107 (1), pp. 168–169. DOI: 10.2307/2152152

Krasner, Stephen D. (1982): Regimes and the Limits of Realism. Regimes as Autonomous Variables. *International Organization* 36 (2), pp. 497–510. DOI: 10.1017/S0020818300019032

Lake, David A. (2010): Rightful Rules. Authority, Order, and the Foundations of Global Governance. *International Studies Quarterly* 54 (3), pp. 587–613. DOI: 10.1111/j.1468-2478.2010.00601.x

March, James G.; Olsen, Johan P. (1998): The Institutional Dynamics of International Political Orders. *International Organization* 52 (4), pp. 943–969. DOI: 10.1162/002081898550699

Mearsheimer, John J. (1994): The False Promise of International Institutions. *International Security* 19 (3), p. 5. DOI: 10.2307/2539078

Morgenthau, Hans Joachim (1948): *Politics Among Nations. The Struggle for Power and Peace*. New York: Alfred A. Knopf.

Okereke, Chukwumerije; Bulkeley, Harriet; Schroeder, Heike (2009): Conceptualizing Climate Governance Beyond the International Regime. *Global Environmental Politics* 9 (1), pp. 58–78. DOI: 10.1162/glep.2009.9.1.58

Rosenau, James N. (1995): Governance in the Twenty-first Century. *Global Governance* 1 (1), pp. 13–43. DOI: 10.1163/19426720-001-01-90000004

Rosenau, James N. (2007): Governing the Ungovernable. The Challenge of a Global Disaggregation of Authority. *Regulation Governance* 1 (1), pp. 88–97. DOI: 10.1111/j.1748-5991.2007.00001.x

Sands, Philippe (2003): *Principles of International Environmental Law*. Cambridge: Cambridge University Press.

Sands, Philippe; Klein, Pierre; Bowett, Derek W. (2009): *Bowett's Law of International Institutions*. 6 ed. London: Sweet & Maxwell.

Scharpf, Fritz W. (1997): *Games Real Actors Play. Actor-Centered Institutionalism in Policy Research*. Boulder, CO: Westview Press.

Schermers, Henry G.; Blokker, Niels (2011): *International Institutional Law. Unity Within Diversity*. 5th rev. ed. Boston, MA: Martinus Nijhoff Publishers.

Schmidt, Brian C. (2013): On the History and Historiography of International Relations. In Walter Carlsnaes, Thomas Risse, Beth A. Simmons (Eds.): *Handbook of International Relations*. Los Angeles, CA: Sage, pp. 3–28.

Scott, Karen N. (2013): Managing Fragmentation Through Governance. International Environmental Law in a Globalised World. In Andrew Byrnes, Mika Hayashi, Christopher Michaelsen (Eds.): *International Law in the New Age of Globalization*. Leiden: Nijhoff, pp. 207–238.

Skubiszewski, Krzysztof (1989): Implied Power of the International Organizations. In Yoram Dinstein (Ed.): *International Law at a Time of Perplexity. Essays in Honour of Shabtai Rosenne. With Assistance of Shabtai Rosenne*. Dordrecht: Nijhoff, pp. 855–868.

Stein, Arthur (1982): Coordination and Collaboration: Regimes in an Anarchic World. *International Organization* 36 (2), pp. 299–324.

Strange, Susan (1982): Cave! Hic Dragones: A Critique of Regime Analysis. *International Organization* 36 (2), pp. 479–496. DOI: 10.1017/S0020818300019020

Ulfstein, Geir (2008): Treaty Bodies. In Daniel Bodansky, Jutta Brunnée, Ellen Hey (Eds.): *The Oxford Handbook of International Environmental Law*. Oxford: Oxford University Press.

UN (1966): Yearbook of the International Law Commission 1966. *Documents of the Second Part of the Seventeenth Session and of the Eighteenth Session Including Reports of the Commission to the General Assembly Volume II*. New York, United Nations Publication. DOI: 10.18356/91cc8d73-en

United Nations. (1993). Part two: Legal activities of the United Nations and related intergovernmental organizations. Chapter VI: Selected legal opinions of the secretariats of the United Nations and related intergovernmental organizations. *United Nations Juridical Yearbook* (pp. 334–458). United Nations.

Vaubel, Roland (1996): Bureaucracy at the IMF and the World Bank. A Comparison of the Evidence. *World Economy* 19 (2), pp. 195–210. DOI: 10.1111/j.1467-9701.1996.tb00672.x

Waltz, Kenneth N. (1979): *Theory of International Politics*. Reading, MA: Addison-Wesley.

Wendt, Alexander (1999): *Social Theory of International Politics*. Cambridge: Cambridge University Press.

Wiersema, Annecoos (2009): The New International Law-Makers? Conferences of the Parties to Multilateral Environmental Agreements. *Michigan Journal of International Law* 31 (1), pp. 231–287.

Williamson, O. E. (1985): *The Economic Institutions of Capitalism*. New York: Free Press.

2 Redefining MEAs as Authoritative International Organisations

The global governance approach observes a disaggregation of authority in the international landscape away from states as the only focal points and towards a plethora of other actors who can attain relevance and become essential players in their own right. This new perspective opens up a broader discussion on the concept of authority, how it should be understood and defined, and precisely to whom it can be applied. This chapter discusses the concept of authority more broadly and applies the notion of relational authority to Multilateral Environmental Agreements (MEAs) specifically. By doing so, it puts forward a conceptualisation of MEAs as International Organisations capable of exercise authority over their member states in their own right. It explains how the Conference of the Parties (COP) becomes a governor in its own right through the performance of specific functions for the governed states through an authority equilibrium. This chapter further presents a list of governance functions relevant for the exercise of authority by MEAs. The understanding of MEAs as authoritative organisations carved out here provides the starting point for an investigation into the phenomenon of informal MEA authority expansion and its specific driving forces undertaken in the following chapters.

Redefining Authority beyond the Nation-State

The disaggregation of authority on the international level that has been described in the previous chapter has changed our understanding of the concept in a way that transcends its original boundaries. It moves away from the traditional focus on formal characteristics as origins of authority and towards a more open, less-formal conceptualisation. Emerging from Rosenau's arguments, several new concepts of authority have emerged that challenge the classical state-centric perspective on world politics and aim to grasp the newly emerging centres of gravity. While these new approaches differ in many ways, they all agree that the old paradigm of IR, which locates authority exclusively with sovereign states, can no longer be maintained (Hickmann 2017, p. 436).

DOI: 10.4324/9781003597681-3

Authority is now a contested concept that has been defined in many ways. The literature on newly emerging forms of authority has received criticism for its analytical inconsistencies and conceptual stretching. Katsikas (2010) goes as far as to say that this broad application of the term authority "stretches its analytical foundations to such lengths that eventually renders it ineffective as an analytical concept, turning it into a blanket concept that can be used to describe everything and anything" (p. 125). Thus, it is important to review the broad landscape of authority concepts and arrive at a clear and specific definition before moving on to apply the concept to the institutions of interest.

Authority used to be exclusively bound to the rational–legal legitimacy of the state. But more abstractly conceived, authority is not bound to the state (Hooghe and Marks 2015). What all instances of authority have in common is that "those who recognise authority defer their own judgement or choice without being necessarily forced or persuaded to do so" (Zürn et al. 2012, p. 86). It is not the force behind a command but the recognition of the one making it a trusted source of orientation that produces adherence. However, this does not necessarily imply blindly following orders, but merely suspending careful elaboration or test of the claims made by that authority (Zürn et al. 2012, p. 86).

Several types of authority can be distinguished. Epistemic authority results from the possession of special knowledge or moral expertise. Individuals or institutions are recognised as an authority if the views they express are recognised as trustworthy. It is not the quality of the argument that is decisive in this case, but the reputation of the person or institution giving it. In modern times, epistemic authorities always compete with one another, and subjects are essentially free to decide whom to believe and, thus, who to consider an authority and who not. In contrast, political authority is exercised by an individual or institution that is assigned the task of authoritative interpretation of facts and norms for the members of a defined group. Here, the prescriptions, rules, and orders made by the one in charge are recognised as collectively binding (Zürn et al. 2012, pp. 86–87). Private authority rests between these two categories. Private actors may gain influence based on their expertise and their acceptance by their audience as epistemic authorities. They may also exert political authority, either for a particular community of (usually non-state) actors for which they are authorised to make binding decisions or based on powers selectively delegated to them by public international institutions (Katsikas 2010; Green 2013b). Andonova (2010), in this context, observes the emergence of hybrid authority through public–private partnerships managed by International Organisations and private actors. And, in an even broader approach, Hall and Biersteker (2002) identify market authority, moral authority, and even illicit authority.

Authority also needs to be distinguished from the closely related concept of autonomy. While the authority of an institution relates to the question of which decisions are made at the collective level, autonomy relates to how these decisions are made. With growing autonomy, the internal constitution of an institution becomes increasingly relevant in shaping and directing decisions (Dörfler and Gehring 2015). Authority and autonomy exist independently of each other. International institutions can be authoritative without being autonomous from powerful states, and they can also have autonomy but not be authoritative (Cooper et al. 2008, p. 505).

In the context of international institutions generally and MEAs specifically, the concept of political authority is most relevant and thus will be specified further. At its core, political authority is rightful or legitimate rule. When political authority is exercised, a governor commands a set of subordinates, the governed, to alter their actions. Command implies that the governor has the right to issue such orders, which, in turn, also implies a duty of the governed to comply (Lake 2007, p. 50). Understood in this way, authority is a form of power. Power has been defined as "the ability of A to get B to do something he would otherwise not do" (Lake 2010, p. 592; Dahl 1957). This is also the case in an authority relationship. What is central for authority, though, is that the subordinate's behaviour is driven not by force but by a feeling of obligation or duty to comply or an acceptance of the legitimacy of the command. Thus, authority is analytically distinct from coercion (Lake 2010, p. 588).

For a long time, scholars of IR have relied on a formal–legal interpretation of political authority rooted in the work of Max Weber (1978). This perspective implies that the ability of the governor to command and the willingness of the governed to comply follows from the lawful position of the office the governor holds. It follows that authority is law, and law is authority. Further, since there is no lawful position above the state, there can be no authority in international politics. It is, from this perspective, always and indisputably a realm of anarchy (Lake 2007, p. 53). Such an understanding of authority might be useful for analysing established domestic hierarchies but is unduly inappropriate for an international setting. If authority derives from law, then law must precede authority. But if authority creates law, then authority must precede lawful office. The formal–legal approach cannot conceive of law without authority or authority without law. Thus, the origins of authority must rest on something other than a formal–legal order. Therefore, it cannot follow that without a formal–legal structure, there can be no authority (Lake 2007, p. 54).

To provide a more useful conceptualisation of authority which resonates more closely with the global governance perspective, Lake (2010) proposes that:

> global governance and its many forms can be understood and unified by a concept of relational authority, which treats authority as a social contract in which a governor provides a political order of value to a community in exchange for compliance by the governed with the rules necessary to produce that order
>
> (p. 589)

Applied to international institutions, relational authority results from an exchange between the institution as the ruler and its member states as the ruled. States confer the right to the institution to exert the restraints on their behaviour necessary to provide a desired order. This resonates with the interactional perspective on international legislatures discussed in the realm of international law as described above (Brunnée 2002), which understands law as resulting from a mutually generative process and breaks open the strict dichotomy between binding and non-binding rules.

From this perspective, authority rests on a bargain between the ruler and the ruled in which the provided order of value by the former is sufficiently beneficial to offset the loss of freedom of the latter. An equilibrium arises in which the institution provides just enough of the political order to gain the member state's compliance with the constraints on behaviour required to sustain this order, and the states comply just enough to induce the institution to provide it. The institution gets a sufficient return on effort to make the provision of order worthwhile, and the states get sufficient order to offset their loss of freedom entailed in consenting to the institution's authority. If the institution extracts too much or provides too little, the member states will withdraw their compliance and the institution's authority evaporates. It follows that authority is contingent on the actions of both the institution and the member states. Thus, relational authority is an equilibrium produced and reproduced through ongoing interactions (Lake 2010, p. 596). It follows that "authority is not law, but a contract" (Lake 2007, p. 54). Here, obligation does not flow from the commands of the ruler but from the consent of the ruled. A ruler only possesses authority if their subordinates acknowledge an obligation or sense of duty to comply with their will. This sense of obligation does not spring from coercion but rather from an interest in and satisfaction with the social order so produced (Lake 2007, p. 55).

Similarly, Zürn et al. (2012, pp. 83–84) describe two layers of legitimate authority. The first layer is the recognition of the authority as functionally necessary to achieve the desired order of value. Therefore, it is granted the competence to make certain decisions and judgments, meaning that institutions have authority when the addressees of their policies recognise that these institutions can make competent judgements and binding decisions. The second layer is the acknowledgement of the rightful exercise of authority in the context of a given stock of normative beliefs in a community. Thus, political authority and rule are legitimate when the norms, rules, and judgements produced are based on shared beliefs about the desired order of value and procedural fairness. In this sense, authority can exist independent of legitimacy but is unlikely to persist if its decision-making procedures and decisions are seen to be regularly and permanently unrightful.

Still, a sense of duty to comply with legitimate authority also implies a right to enforce commands in the event of non-compliance. In an authority relationship, individuals choose whether to comply, but they are bound by the right of the governor to discipline or punish if they choose not to. Lake compares this fact to the speed limit. He argues that many drivers exceed it but also accept the right of the state to issue fines. Ultimately, authority, specifically the right to punish noncompliance, relies on the collective acceptance of the governor's right to rule. The capacity of the governor to enforce in individual cases rests on the collective affirmation and possibly active consent of their subjects. Thus, if an individual denies their obligations, but the larger community still supports the governor's right to punish, the individual can still be bound by their authority. Because a sufficient number of members accept the governor and their decisions as rightful, the governor can enforce their will against individual free riders and even dissidents. The knowledge that a sufficient number of others support the governor's right to enforce can, in turn, be enough to deter potential free riders and dissidents from violating the rules, which renders actual enforcement unnecessary or, at least, unusual. It follows that authority is never a dyadic trait between the governor and individual subjects but derives from a collective conferral of rights. Recognising this resolves the apparent contradiction that, from the collective perspective, compliance with authority is voluntary, but from the standpoint of individual members, compliance is mandatory. It is individuals who obligate themselves to follow the commands, but they still choose collectively whether to accept the governor's authority or not (Lake 2010, p. 592).

The relational understanding of authority represents a move away from a formal–legal conception. It leaves behind the focus on formal characteristics as sources of authority and is instead rooted in the functions an institution performs for its members. Relational authority is rooted in the ability of any type of institution to make decisions on rules and procedures which are accepted as collectively binding by all members because they are expected to produce the desired order of value the institution was created to produce:

> international institutions have authority when states recognise, in principle or in practice, their ability to make legally binding decisions on matters relating to a state's domestic jurisdiction.
>
> (Cooper et al. 2008, p. 505)

These legally binding decisions are, in essence, governance functions an institution is mandated to perform in order to produce a collective good. Based on the functional logic of a relational concept of authority as outlined above, Zürn et al. state that:

> International institutions exercise authority in that they successfully claim the right to perform regulatory functions like the formulation of rules and rule monitoring or enforcement
>
> (2012, p. 70)

Nothing in this relational conception of authority requires a state or state-like entity (Lake 2010, p. 594). Relational authority can exist beyond the nation-state and does not even require strong and independent International Organisations. This definition is not limited to institutions in which functions are delegated to a separate agent, like the bureaucracies of traditional IOs, but explicitly includes institutions where sovereignty is pooled in a collective body.

This new understanding of authority beyond the nation-state can be applied to MEAs without restrictions and allows for a conceptualisation of MEAs as separate entities capable of exercising authority in global governance. As the following chapters will show, if the concept of relational authority is taken seriously, MEAs are inevitably understood as governors in their own right, capable of exercising authority through the performance of regulatory functions by adopting legally binding decisions. This represents a significant departure from the more established understandings of MEAs discussed in earlier chapters and provides a sound basis for investigating the question of the true nature of MEA authority and the driving forces of its expansion.

How MEAs Exercise Authority in Global Governance

Governance has been defined as the exercise of authority. Based on the definition of relational authority outlined in the previous chapter, we now reach a conceptualisation of MEAs as separate entities exercising authority on the international level. Relational authority explicitly encompasses not only traditional IOs in which authority is delegated to large-scale independent bureaucracies but also institutions in which authority is pooled, as is the case in the COPs of MEAs. Pooling can essentially be understood as a situation in which states delegate functions to a collective of themselves (Cooper et al. 2008; Bradley and Kelley 2008). Under pooling, states remain the central decision-makers. Still, they have to abide by the outcome produced collectively, "thereby subjecting a state's policies to an external actor without granting the individual state any veto power" (Cooper et al. 2008, p. 506). Recently, indeed, following a similar line of argumentation, the point has been made that member-dominated IOs in which sovereignty is pooled, even including MEAs specifically, can become actors in their own right if they also attain a level of autonomy (Gehring and Urbanski 2023; Gehring and Spielmann 2023).

Traditional approaches in global governance continue to see states as supreme in principle, and any authority possessed by others is understood as merely delegated by them. In this view, states may allow others to exercise authority for a time. Still, they always reserve the right and ability to retract that authority and impose their own legitimate control instead. But, if the relational concept of authority is taken seriously, "even sovereign states have no special status" (Lake 2010, p. 599). States negotiate a realm of authority, but they are also limited by other authorities that are themselves rightful. Global authorities do not exist at the sufferance of states or simply in the interstices of state power but are themselves independent authorities through their own social contracts. It follows that "Global governance is not delegated from states but is as real as the authority possessed by any state" (Lake 2010, p. 599).

In this context, the concept of incomplete contracts, originating from transaction–cost–economics (Milgrom and Roberts 1990), can be applied to MEAs. The central premise is that a contract which calls for the future delivery of a good or service, the provision of capital, or the future performance of work can never specify exactly what actions are to be taken and what payments are to be made in all possible future contingencies. This is true for several reasons. First, parties to a contract can never perfectly anticipate all contingencies they may encounter in relation to the agreed treaty provisions. Second, even for those circumstances that can be anticipated, it is often more economical to respond to them when the need arises rather than to plan in advance for every foreseeable instance. Third, it is almost impossible to write completely unambiguous contracts. Including too many fine distinctions simply increases the likelihood of emerging events falling into areas of ambiguity or overlap, leading to disagreements that must be resolved. Finally, contracting partners might not share all information that might become relevant for the cooperation, and complete contracts cannot be based on information that only one contracting party will have. Thus, contracting parties:

> content themselves with an agreement that frames their relationship – that is, one that fixes general performance expectations, provides procedures to govern decision-making in situations where the contract is not explicit, and outlines how to adjudicate disputes when they arise.
>
> (Milgrom and Roberts 1990, p. 62)

Such incomplete contracts relieve actors from having to decide everything at once and leave room for adaptation to changing circumstances by transferring some decisions to later stages of their cooperation project. It follows that incomplete contracting always involves two stages of decision-making, namely the decision to establish (or adapt) an institutional framework and (secondary) decisions adopted within this framework. Secondary decisions can but need not be delegated to a specifically created agent. But, where uncertainty is great and future decision-making is expected to be time-consuming and complex, parties may choose to create a specialised body they can delegate these tasks to (Pollack 1997, pp. 103–104). Then, the incomplete contract itself amounts to the "constitution" of an emerging institution in which procedures for secondary decision-making are set down. Consequently, the rights and obligations resulting from the cooperation project continuously unfold through the stream of secondary decisions. These are guided by the framework of the original treaty but cannot be entirely anticipated therein and will necessarily include a significant degree of independence.

Like the parties to any contract, it is difficult for states to anticipate and provide for all possible contingencies of a cooperation project on the international level. While the domestic legal system can supply missing terms for an incomplete contract on the national level, the international context is relatively thin in this regard. Here, the creation of stable organisational structures for elaborating rules, standards, and specifications based on an incomplete contract is particularly relevant in addressing problems of transaction costs and opportunism (Abbott and Snidal 1998, p. 15). The concept of incomplete contracting resonates particularly well with the structure of MEAs, their skeletal founding treaties specifying only basic aims and procedures, and their extensive institutional machinery creating a continuous stream of secondary decisions to elaborate the specifics of the cooperation project and render it fully functional. When the constitutional treaty is signed, a number of governance functions are pooled inside the MEA. The individual states sacrifice unilateral control over these functions; they can now only be employed by collective decision. These functions are subsequently exercised by the MEA through secondary decision-making. These secondary decisions become legally binding for all individual members. The MEA exercises authority.

As a consequence, the authority of the collective is as strong and valuable as that of each individual state. Thus, member state control over the actions of MEAs might not be as tight as it seems. The collective of member states forms a separate entity that makes decisions inside a specific institutional context which follows its own logic. An equilibrium emerges in which the MEA provides environmental protection in a specific field (the desired order value) through collectively binding decisions, and the group of individual member states provides their compliance with these decisions to produce the desired environmental protection. Figure 2.1 illustrates this equilibrium relationship.

Through this relational understanding of MEA authority, International Relations theory now reaches a similar conclusion than scholars of International Law have long considered. MEAs are, indeed, "virtually identical to international organisations" (Brunnée 2002, p. 16).

The specific governance functions through which an MEA exercises its authority can be roughly categorised along the stages of the regulatory process, starting with the negotiation of rules and regulations, followed by the implementation of these rules, the collective appraisal of member state behaviour and finally, the handling of non-compliance (Zürn et al. 2012; Abbott and Snidal 2009; Gehring 2008). Many functions discussed in functional theories of institutional design and the principal-agent literature that can be delegated to non-state agents are also relevant in the context of horizontally structured MEAs (Bradley and Kelley 2008; Brown 2010; Pollack 1997; Epstein and O'Halloran 1999).

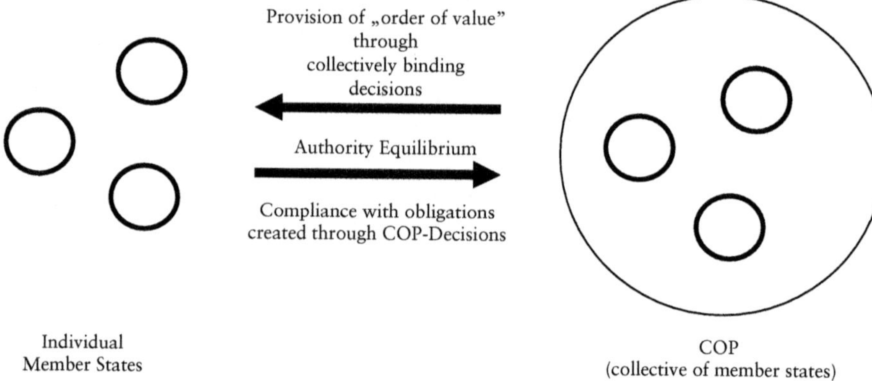

Figure 2.1 Relational authority of MEAs.

Rulemaking

MEAs adopt specific rules subsequent to the initial framework of the incomplete contract. Frequently, the constitutional treaty needs to be fleshed out to become applicable in a meaningful way and effective in creating the desired order of value. Such rules frequently define the operation of internal institutional processes but have important external effects on their addressees (Schermers and Blokker 2011, pp. 761–766). Occasionally, MEAs have even introduced major changes to the regulatory scheme. For example, the COP of the Basel Convention of the Transboundary Movement of Hazardous Wastes adopted a decision that replaced the original regulatory control scheme of the convention according to a prior informed consent procedure with a complete ban on exports of hazardous wastes from OECD countries to non-OECD countries (Kummer 1998, p. 229).

The authority to make rules implies particularly high costs for the sovereignty of member states. After all, the rules an institution makes have the potential to significantly affect member states' interests (Green and Colgan 2013). Even though some rules made might require ratification by member states, the institutional decision provides a proposal that cannot be amended individually but only accepted or rejected. A firm grasp on this resource turns an MEA into an issue-specific global legislature (Brunnée 2002, p. 4).

Rule-making power can vary in the level of obligation these regulations attain (Abbott and Snidal 2000, p. 424). While the ability to make rules with a higher level of obligation, as well as the ability to regulate a higher number of areas, certainly means a stronger grasp on this resource, too strict a distinction between hard and soft law might have few practical effects.

Authority does, after all, rely on consent and not on force. Since members desire the order of value that these rules create, the benefits of observing them clearly outweigh the disadvantages of ignoring them. Thus, they will likely take them seriously either way.

Implementation Decisions

The next step along the regulatory process is the implementation of the established rules. Many institutions require regular decision-making on a day-to-day or case-by-case basis. Thus, a function MEAs regularly perform is the adoption of specific implementation decisions, e.g., funding and lending decisions or classifications of species or substances within a regulatory scheme. For example, the International Whaling Commission, established under the International Convention for the Regulation of Whaling, regularly decides on the maximum allowed catch and other measures intended to protect populations. Collective control over funding decisions entails opportunity costs in terms of what other policies a member might have been able to create on their own instead (Bradley and Kelley 2008, p. 28). While these decisions implement the constitutional treaty without adding new rules, they are made at the collective level and have consequences for the individual states, generally without the need for ratification.

Collective Re-Delegation

A specific form of implementation decisions is the – formal or informal – power of an institution to re-delegate some of its authority to other public or private entities to be fulfilled under specific conditions and according to established criteria set at the collective level. Re-delegation can also entail the creation of new bodies that emanate from the original international body. Some forms of authority may be more frequently re-delegated than others. Implementation delegation is the most common, while re-delegation of regulatory authority is rare (Bradley and Kelley 2008, p. 17). This function is likely to reflect attempts to minimise negative effects and inconsistencies and exploit the positive effects of institutional overlap in regime complexes (Gehring and Faude 2014; Raustiala and Victor 2004). Since their legal capability to conclude formal treaties under International Law remains somewhat undetermined, agreements may appear in other forms, such as memoranda of understanding. This includes, for example, the coordination of regulatory or reporting activities or the conditioning of funding organised elsewhere. For example, the Biodiversity Convention employed this resource when it outsourced its funding mechanism to the Global Environment Facility, which approves and manages projects according to criteria provided for by the convention.

Monitoring and Transparency

Once rules are successfully implemented, the next step is the collective monitoring of individual member state compliance with these rules. MEAs generally include measures to produce transparency, verify compliance, and detect cheating. This becomes relevant for the authority of MEAs when the institution becomes more than a "clearing house for self-reporting" (Brown 2010, p. 146) and has some autonomy in selecting time, place, or intrusiveness of monitoring activities or performing analysis on acquired information. It is particularly strong if states can make a legal declaration of (non-)compliance (Brown 2010, p. 147).

Strategies can range from voluntary reporting standards to mandatory onsite inspections. They can be carried out either by a standing body or on an ad hoc basis. In some cases, the institution might only be mandated to collect and distribute information, while others might have the authority to determine and declare whether a state is in compliance or not. Environmental treaties typically create specific bodies to which members become obligated to submit regular reports (Bradley and Kelley 2008, p. 12). Here, failure to report generally indicates improper behaviour in itself (Abbott and Snidal 1998). Monitoring mechanisms are likely to be particularly effective if they are open to information from non-governmental sources such as non-governmental organisations (Tallberg et al. 2013). The most prominent example is the Paris Agreement's enhanced transparency framework, which includes extensive mandatory reporting for all parties as well as a technical and multilateral review process.

Enforcement

States might decide not to comply with the rules of an MEA. While compliance is collectively voluntary, states might individually cheat when they dissembled their interests *ex ante* or their interests change *ex post* (Brown 2010, p. 146). States might have a considerable incentive to take a free ride, or they might stop following their own obligations if they suspect others to neglect theirs (Tallberg 2002, p. 612). Where incentives to free ride prevail, as is the case in nearly any environmental treaty, effective mechanisms to prevent non-compliance are essential. Non-compliance always threatens to undermine established cooperation (Gehring 1994, p. 314). Enforcement measures may be necessary to align some actors' interests to make cheating less advantageous (Brown 2010, p. 147). They seek to influence an actor's decision-making process by increasing the costs of non-compliance so that it becomes more expensive than compliance (Bodansky 2010a, pp. 235–236).

Enforcement typically refers to the application of rewards and penalties (Abbott and Snidal 2009, p. 65). While international institutions do not possess jurisdiction to enforce rules in the same way as states do, they may produce a softer version of pressure to comply through sanctions, shaming or fines (Zürn et al. 2012, p. 87). Thus, monitoring strategies are not only a preparation for the handling of non-compliance but may itself be designed to induce compliance with international obligations (Bradley and Kelley 2008, pp. 12–14). The same forums in which ambiguities can be explained and resolved can also be used to bring pressure on transgressor states. This "mobilisation of shame" is not only targeted at creating guilt among states but more towards enhancing reputational and other incentives to abide by their commitments. Such practices pressure states to change their behaviour both by impairing their international standing and by empowering private groups to pressure national governments, thus increasing "audience costs", which might lead to domestic audiences questioning whether the leadership is successful or not in foreign policy (Fearon 1994; Abbott and Snidal 1998, p. 26).

Although enforcement is generally difficult in global governance, where member states can simply leave institutions if punishment might lead to costs exceeding the benefits, several more tangible forms of enforcement might be available. MEAs might be able to deprive non-compliant members of valuable benefits, e.g., participation in a particular regulatory scheme like emissions trading under the Kyoto Protocol. Other forms of enforcement might be the suspension of voting rights or membership. Finally, MEAs might even be able to exercise stronger forms of enforcement in the form of sanctions (Schermers and Blokker 2011, pp. 913–984; Bradley and Kelley 2008, p. 13; Brown 2010, p. 147). A rare example is the exceptionally strong enforcement branch of the Kyoto Protocol, which was able to impose penalties for subsequent commitment periods if a party failed to reach its emission reduction target (see section "The Kyoto Enforcement Branch" in Chapter 5).

Compliance Management

It has been argued that the more common sources of non-compliance are not deliberate misconduct but ambiguities in treaty language or limitations of the capacity of parties to carry out their obligations in a sufficient manner. Under this assumption, addressing non-compliance effectively requires a strategy of integrated, active management instead of punishment through sanctions (Chayes et al. 2000; Chayes and Chayes 1995). While the enforcement approach would likely result in a loss of eligibility for support, this would, under the circumstances described above, exacerbate rather than solve non-compliance. Thus, the managerial approach to non-compliance focuses on prevention instead of retribution and aims to encourage and facilitate compliance. Compliance management has become more relevant in international environmental protection also because, in many cases, once the harm is done, it can hardly be reversed. Therefore, it is better to promote compliance from the beginning and bring states back on track as fast as possible instead of simply punishing misconduct and potentially discouraging further participation. In this area, MEAs might gain authority in the form of the establishment of a specialised compliance committee, which is mandated to manage instances of non-compliance in an amicable, non-punitive way. Rather than recommend sanctions, such a committee can negotiate action plans with the non-complying party, including possible subsidies and support, and focus on resolving compliance issues through mutual consultation and deliberation, promoting transparency and building national capacity (Bodansky 2010a, p. 236). The most prominent example of this is the compliance committee of the Montreal Protocol, which was the first to introduce this approach (see section "Compliance Management" in Chapter 4).

The managerial approach to non-compliance is not incompatible with the enforcement model. While the managerial model downplays the importance of sanctions in favour of a facilitative approach, it does not completely preclude a role for them (Bodansky 2010a, p. 238). Different reasons for non-compliance might be present at any time and require different responses.

Implementation Assistance

Taking a managerial view on non-compliance further, MEAs generally include extensive mechanisms for the provision of assistance to those that might struggle to fulfil their obligations. States generally operate under a sense of obligation to conform to international rules, and they have an interest in the stability of the order of value that arises from collective compliance. But, some might lack the necessary capacities to thoroughly fulfil their commitments. Capacity problems cannot be solved through strict enforcement measures. Instead, the institution may decide to provide financial or technical assistance (Abbott and Snidal 1998, p. 26; Chayes et al. 2000; Chayes and Chayes 1993). The creation of the Interim Multilateral Fund of the Montreal Protocol (see section "The Creation of the Interim Multilateral Fund" in Chapter 4) is an impressive example, as well as the extensive channels under the Paris Agreement to share technical knowledge.

References

Abbott, Kenneth W.; Snidal, Duncan (1998): Why States Act through Formal International Organizations. *Journal of Conflict Resolution* 42 (1), pp. 3–32. DOI: 10.1177/0022002798042001001

Abbott, Kenneth W.; Snidal, Duncan (2000): Hard and Soft Law in International Governance. *International Organization* 54 (3), pp. 421–456. DOI: 10.1162/002081800551280

Abbott, Kenneth W.; Snidal, Duncan (2009): Chapter Two: The Governance Triangle. Regulatory Standards Institutions and the Shadow of the State. In Ngaire Woods, Walter Mattli (Eds.): *The Politics of Global Regulation*. Princeton, NJ: Princeton University Press, pp. 44–88.

Andonova, Liliana B. (2010): Public-Private Partnerships for the Earth. Politics and Patterns of Hybrid Authority in the Multilateral System. *Global Environmental Politics* 10 (2), pp. 25–53. DOI: 10.1162/glep.2010.10.2.25

Bodansky, Daniel (2010a): *The Art and Craft of International Environmental Law*. Cambridge, MA: Harvard University Press.

Bradley, Curtis A.; Kelley, Judith (2008): The Concept of International Delegation. *Law and Contemporary Problems* 71 (1), pp. 1–36.

Brown, Robert L. (2010): Measuring Delegation. *The Review of International Organizations* 5 (2), pp. 141–175. DOI: 10.1007/s11558-009-9076-3

Brunnée, Jutta (2002): COPing with Consent. Law-Making Under Multilateral Environmental Agreements. *Leiden Journal of International Law* 15 (1), pp. 1–52. DOI: 10.1017/S0922156502000018

Chayes, Abraham; Chayes, Antonia Handler (1993): On Compliance. *International Organization*, 47(2), pp. 175–205.

Chayes, Abram; Chayes, Antonia Handler (1995): The New Sovereignty. Compliance with International Regulatory Agreements. Cambridge, MA: Harvard University Press.

Chayes, Abram; Chayes, Antonia Handler; Mitchell, Ronald B. (2000): Managing Compliance. A Comparative Perspective. In E. B. Weiss, H. K. Jacobson (Eds.): *Engaging Countries. Strengthening Compliance with International Environmental Accords*. Cambridge, MA: MIT Press, pp. 39–62.

Cooper, Scott; Hawkins, Darren; Jacoby, Wade; Nielson, Daniel (2008): Yielding Sovereignty to International Institutions. Bringing System Structure Back In. *International Studies Review* 10 (3), pp. 501–524. DOI: 10.1111/j.1468-2486.2008.00802.x

Dahl, Robert A. (1957): The Concept of Power. *Systems Research and Behavioral Science* 2 (3), pp. 201–215. DOI: 10.1002/bs.3830020303

Dörfler, Thomas; Gehring, Thomas (2015): Wie internationale Organisationen durch die Strukturierung von Entscheidungsprozessen Autonomie gewinnen. Der Weltsicherheitsrat und seine Sanktionsausschüsse als System funktionaler Ausdifferenzierung. In Andrea Liese, Eugénia da Conceiçao-Heldt, Martin Koch (Eds.): *Internationale Organisationen. Autonomie, Politisierung, interorganisationale Beziehungen und Wandel* . Baden-Baden: Nomos, pp. 54–80.

Epstein, David; O'Halloran, Sharyn (1999): *Delegating Powers*. Cambridge: Cambridge University Press.

Fearon, James D. (1994): Domestic Political Audiences and the Escalation of International Disputes. *American Political Science Review* 88 (3), pp. 577–592. DOI: 10.2307/2944796

Gehring, Thomas (1994): *Dynamic International Regimes. Institutions for International Environmental Governance*. Frankfurt am Main: Lang.

Gehring, Thomas (2008): Treaty-Making and Treaty Evolution. In Daniel Bodansky, Jutta Brunnée, Ellen Hey (Eds.): *The Oxford Handbook of International Environmental Law*. Oxford: Oxford University Press.

Gehring, Thomas; Faude, Benjamin (2014): A Theory of Emerging Order Within Institutional Complexes. How Competition Among Regulatory International Institutions Leads to Institutional Adaptation and Division of Labor. *The Review of International Organizations* 9 (4), pp. 471–498. DOI: 10.1007/s11558-014-9197-1

Gehring, Thomas; Spielmann, Linda (2023): The Treaty Management Organization Established Under the UNFCCC and the Paris Agreement: An International Actor in Its Own Right? *International Environmental Agreements*. DOI: 10.1007/s10784-023-09611-z

Gehring, Thomas; Urbanski, Kevin (2023): Member-Dominated International Organizations as Actors: A Bottom-Up Theory of Corporate Agency. *International Theory* 15 (1), pp. 129–153. DOI: 10.1017/S1752971922000069

Green, Jessica F. (2013b): *Rethinking Private Authority. Agents and Entrepreneurs in Global Environmental Governance*. Princeton, NJ: Princeton University Press.

Green, Jessica F.; Colgan, Jeff (2013): Protecting Sovereignty, Protecting the Planet. State Delegation to International Organizations and Private Actors in Environmental Politics. *Governance* 26 (3), pp. 473–497. DOI: 10.1111/j.1468-0491.2012.01607.x

Hall, Rodney Bruce; Biersteker, Thomas J. (2002): The Emergence of Private Authority in the International System. In Rodney Bruce Hall, Thomas J. Biersteker (Eds.): *The Emergence of Private Authority in Global Governance*. Cambridge: Cambridge University Press (85), pp. 3–22.

Hickmann, Thomas (2017): The Reconfiguration of Authority in Global Climate Governance. *International Studies Review* 19 (3), pp. 430–451. DOI: 10.1093/isr/vix037

Hooghe, Liesbet; Marks, Gary (2015): Delegation and Pooling in International Organizations. *The Review of International Organizations* 10 (3), pp. 305–328. DOI: 10.1007/s11558-014-9194-4

Katsikas, Dimitrios (2010): Non-State Authority and Global Governance. *Review of International Studies* 36 (S1), pp. 113–135. DOI: 10.1017/S0260210510000793

Kummer, Katharina (1998): The Basel Convention: Ten Years On. *Review of European Community & International Environmental Law* 7 (3), pp. 227–236. DOI: 10.1111/1467-9388.00154

Lake, David A. (2007): Escape from the State of Nature. Authority and Hierarchy in World Politics. *International Security* 32 (1), pp. 47–79. DOI: 10.1162/isec.2007.32.1.47

Lake, David A. (2010): Rightful Rules. Authority, Order, and the Foundations of Global Governance. *International Studies Quarterly* 54 (3), pp. 587–613. DOI: 10.1111/j.1468-2478.2010.00601.x

Milgrom, Paul; Roberts, John (1990): Bargaining Costs, Influence Costs, and the Organization of Economic Activity. In James E. Alt, Kenneth A. Shepsle (Eds.): *Perspectives on Positive Political Economy*. Cambridge: Cambridge University Press, pp. 57–89.

Pollack, Mark A. (1997): Delegation, Agency, and Agenda Setting in the European Community. *International Organization* 51 (1), pp. 99–134.

Raustiala, Kal; Victor, David G. (2004): The Regime Complex for Plant Genetic Resources. *International Organization* 58 (2), p. 347. DOI: 10.1017/S0020818304582036

Schermers, Henry G.; Blokker, Niels (2011): *International Institutional Law. Unity within Diversity*. 5th rev. ed. Boston, MA: Martinus Nijhoff Publishers.

Tallberg, Jonas (2002): Paths to Compliance. Enforcement, Management, and the European Union. *International Organization* 56 (3), pp. 609–643. DOI: 10.1162/002081802760199908

Tallberg, Jonas; Sommerer, Thomas; Squatrito, Theresa; Jonsson, Christer (2013): *The Opening Up of International Organizations*. Cambridge: Cambridge University Press.

Weber, Max (Ed.) (1978): *Economy and Society. An Outline of Interpretive Sociology*. New York: University of California Press.

Zürn, Michael; Binder, Martin; Ecker-Ehrhardt, Matthias (2012): International Authority and Its Politicization. *International Theory* 4 (1), pp. 69–106. DOI: 10.1017/S1752971912000012

3 A Theoretical Framework for Informal MEA Authority Expansion

Multilateral Environmental Agreements (MEAs) have been firmly established as possessing authority of their own and becoming separate entities capable of exerting authority on the international level in the previous chapter. Now, we observe not only that MEAs exercise authority at all but also that they further expand their authority over time, claiming new functions as they mature. Conceptualising them as International Organisations (IOs) allows for an investigation into the driving forces which make these expansions possible. Authority, as defined here, is inherently dynamic and subject to constant negotiation and re-negotiation. Thus, significant shifts may become possible under the right circumstances, even without formal delegation. This chapter carves out a theoretical framework for the analysis of informal MEA authority expansion. First, theoretical foundations based in the dynamic nature of relational authority and an informal interpretation of the principal-agent (PA)-approach are laid down. Second, two driving forces of informal authority expansion are proposed by fleshing out this groundwork through a combination with the neofunctional logic of spill-over as one of the most prominent approaches to the study of institutional expansion, usually reserved for the European context. This chapter also puts forward observable implications for both driving forces and methodological considerations for how this framework will be put into practice.

Foundations of Informal MEA Authority Expansion

Lake (2010, p. 589) describes authority as "almost a living thing". Understood in the way employed here, as a contract formalising an exchange relationship, authority can never be static but is inherently dynamic. The understanding that authority equilibria are continuously reproduced through the interaction of both players implies that significant shifts are possible without a formal push. States might be enmeshed in processes, which ultimately lead to the collective engaging in the performance of functions which are not explicitly provided for in the constitutional treaty. Following this argumentation, the rules underlying an authority relationship will be more or less biased towards

DOI: 10.4324/9781003597681-4

the governor and will always be contested and open for re-negotiation by both sides. The MEA might claim additional functions in order to effectively provide the desired order of value, which in turn means that states will have to comply with additional rules and procedures to keep the balance – thereby shifting the authority equilibrium towards increasing MEA authority and decreasing state control.

The possibility of such shifts has, in principle, been recognised in the field of International Law. As such, Wiersema (2009, p. 271) describes that:

> To the extent that consensus-based COP activity enriches the original commitments of the parties, it contributes to the development of specialised regimes with thick legal obligations that are adaptable, shifting over time as the parties develop them through COPs and their subsidiary organs.

Since compliance with MEA rules is only collectively voluntary but individually mandatory, the individual member states have to accept these claims and comply with the additional obligations if they do not want to be punished for non-compliance or even lose the benefit of membership. Institutions are entitled to make decisions which may go against the interest of some member states as long as they are true to a shared understanding of the common good they established the institution to achieve (Zürn et al. 2012; Cooper et al. 2008, p. 505). Thus, in some situations, it might be rational for states to accept a certain loss of control, even if it goes beyond their individual interest.

In fact, MEAs are designed to expand. From a functionalist perspective, institutions exist in the form they do because they fulfil a specific function. Such institutions either "must be products of rational design, or they emerge over time through mechanisms of institutional enhancement" (Pierson 2000, p. 477). In a way, both of these are true for MEAs. They are consciously designed to develop into a form that serves their purpose increasingly well over time. As incomplete contracts, they are set up to be highly dynamic and independently carry out the evolution of substantive regulations inside a specified framework over time. Rules and procedures are frequently tightened with growing scientific and technological knowledge and the gradual emergence of suitable strategies. When becoming a member of the MEA, states commit themselves not only to the existing obligations but also to the procedures which will tighten obligations over time (Gehring 2008; Brunnée 2004; Werksman 2014).

Expansions of institutional authority have been widely theorised in the context of IOs. Some go as far as to compare them to Frankenstein's Monster, created by their masters but having slipped their restraints and now running amok, terrorising the global countryside (Hawkins et al. 2006, p. 4), European Integration being the most prominent but not the only example (Heldt and Schmidtke 2017; Mahoney and Thelen 2009). These considerations have not branched out to more informal forms of cooperation like MEAs. However, if MEAs are recognised as separate entities which are more similar to organisations than regimes, the possibility that similar mechanisms are taking place inside MEAs needs to be considered.

The approach most commonly employed to analyse situations in which IOs act separately and expand their own capacities is the PA-approach in the tradition of rational choice institutionalism. As discussed in Chapter 1, applying this approach to MEAs is not easily possible since it focuses on hierarchical authority relationships between states and the large-scale bureaucracies of IOs, resulting specifically from delegation as opposed to pooling. Newer advancements of the approach, however, provide a starting point to expand its scope to include more informal forms of cooperation like MEAs.

The concept of Informal IO Empowerment (IOE) described by Heldt and Schmidtke (2017) provides an advancement of the PA-approach, which opens up the idea in a way that makes a transfer to the processes expected to take place inside MEAs more plausible. Arguing that the exclusive focus on formal components does not withstand empirical observation, they build on the shortcomings of the PA-approach and organisational perspectives on IO bureaucracies in this regard. IOE refers to the organisational process of transferring specific functions to IOs, which evolves over time and gradually shapes their tasks, scope, and capabilities. These processes do not only take the form of formal changes through treaty amendments they also come in more subtle, incremental modes, which nevertheless have the potential to substantively shape IOs' power over time. Informal IOE occurs when new tasks are added to an IO's portfolio, when the issue areas in which IO tasks are performed are extended, and when staff and financial capabilities increase without changing the formal contract. In contrast to formal reform, informal IOE takes the form of more incremental and subtle changes in the interpretation and application of an IO's formal mandate both by member states and international bureaucrats.

The concept originates from a critical perspective on historical institutionalism and path-dependence, which recognises that institutional change does not always result from external critical junctures but can also be generated endogenously and advance in a more gradual way (Thelen 2003; Thelen and Streeck 2005; Mahoney and Thelen 2009). Theories of institutional development mostly locate significant change in historical ruptures, opening a path for abrupt change and discontinuity. However, in reality, transformative change can often be found to result from an accumulation of gradual and incremental change and is often endogenously produced by the very behaviour an institution itself generates. In these cases, significant change emanates from inherent ambiguities and gaps that exist by design or emerge over time between formal institutions and their actual implementation or enforcement. These gaps can become key sites of political contestation over the form, functions, and salience of specific institutions. In these instances, analytical frameworks that take the absence of disruption as sufficient evidence of institutional continuity clearly miss the point (Thelen and Streeck 2005, pp. 18–19).

A central mechanism of IOE is conversion. Conversion can be seen as a more gradual and indirect version of the classic agency slack. It occurs when rules remain formally the same but are interpreted and implemented in new ways. This mechanism is pushed forward by actors who actively exploit the inherent ambiguities of institutional rules. Through strategic re-deployment of these rules, they convert the institution to new goals, functions, or purposes (Heldt and Schmidtke 2017, p. 54; Mahoney and Thelen 2009, p. 16). Conversion occurs under high levels of institutional discretion in the interpretation and enforcement of rules and where rules are ambiguous enough to permit different (often starkly contrasting) interpretations. Conversion starts with political contestation over what functions and purposes an existing institution should serve, made possible by the gaps that exist by design or emerge over time between institutionalised rules and their local enactment. Policymakers respond by deploying existing institutional resources to new ends. Alternatively, where power relations change such that actors who were not involved in the original design of an institution and whose participation in it may not have been reckoned with they might take over and turn it to new ends (Thelen and Streeck 2005, pp. 26–29, 31).

A precondition for conversion is the ambiguity of institutional rules. In practice, institutional ambiguity is present in a variety of forms. One of the more common forms, however, is that of an institution with multiple, non-complementary logics or goals. Institutional structures induce conversion by highlighting two competing logics; since two goals coexist, the implementation can fluctuate between the two depending on which party is in power (Rocco and Thurston 2014, p. 44). These gaps and ambiguities create different types of pressures which require that decisions be made to resolve them – some of which might ask for the transfer of additional authority resources to the collective level.

The transfer of this approach to the study of MEAs seems fitting to investigate the possibility of informal expansion. The general concepts are deliberately formulated in a relatively general way so that they can travel to different contexts. Even though they explicitly employ a regime perspective, Thelen and Streeck (2005, p. 16) state that:

> what an institution is, is defined by the continuous interaction between rule makers and rule takers during whichever new interpretations of the rule will be discovered, invented, suggested, rejected, or for the time being, adopted,

and institutions are conceived as systems of social interaction under formalised normative control. They are continuously created and recreated by a greater number of actors with divergent interests, varying normative commitments, different powers, and limited cognition. This perspective fits very well with the conceptualisation of MEAs as incomplete contracts and as collective entities exercising relational authority over their member states. The generally broad discretion MEAs enjoy in interpreting the original incomplete contract provides favourable conditions for conversion to take place. This opens up the possibility for MEAs to go beyond their original mandate without a change in the original contract.

As incomplete contracts in highly dynamic environments, MEAs are likely to regularly face problems for which existing rules provide no one clear answer on how they should be solved. Then, pressures to expand the boundaries of the organisation might arise. Generally, two possibilities to react are available (see Figure 3.1). First, member states can negotiate a treaty amendment which formally delegates an additional function to the MEA. In most cases, however, formal amendments require ratification by a specified number of individual member states to enter into force. This process can be tedious and complicated – too much so to choose this path every time an issue needs resolving. Thus, the second possibility is much more commonly employed. MEAs can also handle such a situation on the inside by adopting COP decisions, which become immediately binding on all member states. As outlined earlier, MEAs are deliberately designed to be particularly adaptable and continuously expand their regulatory framework to fit the unique and ever-evolving challenges of environmental protection.

The boundaries of this second path might be surprisingly unpredictable, although not completely arbitrary. If a problem arises, accessible paths for resolution are defined by the rules of the original treaty. The expansion does not happen in vacuo but through the conversion of what is already there. New decisions always have to be connected to previous decisions, which ultimately lead back to the original founding treaty of the MEA. When becoming a party to an MEA, states have, in advance, accepted all these decisions that can result from the gaps and ambiguities of the incomplete contract, but also the guidance and guardrails it nevertheless provides. This is similar to a social systems perspective conceptualising IOs as cognitively open but operatively closed, self-referential systems (Koch 2009, p. 438; Gehring 2002; Luhmann 2021, 1987).

This chapter has shown that MEAs are not only highly dynamic, independently relevant organisations which are capable of exerting authority over their members but also that they can expand this authority without

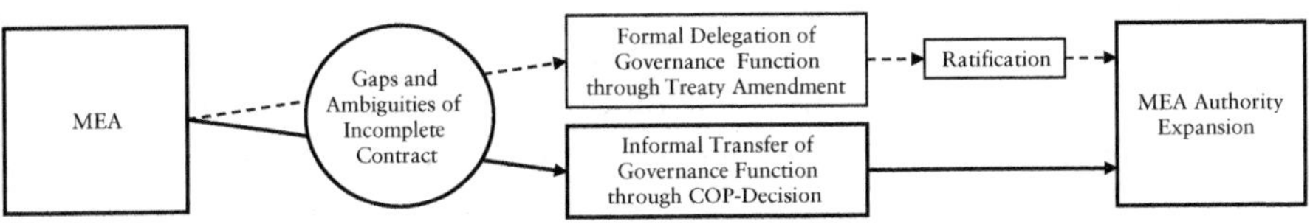

Figure 3.1 Formal and informal MEA authority expansion.

formal changes to the underlying contract. Instead, they can attain new functions through the conversion of ambiguous and incomplete rules in reaction to newly arising and thus far uncovered situations. The question that immediately follows is how these expansions are triggered and what they can look like in practice. The remainder of this chapter carves out a pathway to answering this question by proposing two driving forces and laying out a framework for the identification of these driving forces in specific cases of informal authority expansion.

Theorising Driving Forces of Informal Authority Expansion

The theory which has probably engaged most with driving forces of authority expansions of an IO is Neofunctionalism (Haas 1958, 1961; Lindberg 1963; Lindberg and Scheingold 1970). The European Union (EU) is, after all, the most prominent example of an organisation continuously expanding its authority beyond its original boundaries. One of the pioneers of Neofunctionalism, Ernst Haas, defined European integration as:

> the process whereby political actors in several distinct national settings are persuaded to shift their loyalties, expectations and political activities toward a new centre, whose institutions possess or demand jurisdiction over the pre-existing national states.
>
> (1958, p. 16)

As opposed to the second "grand theory" of European Integration, Liberal Intergovernmentalism (Moravcsik 1998, 1993), which focuses on state bargaining and thus relates more to the classic regime perspective, Neofunctionalism describes a gradual view on the expansion of EU competencies, which is in line with the perspective on MEAs employed here. Neofunctionalism emphasises that integration is primarily driven by incremental decision-making and is often the result of unintended consequences and imperfect knowledge. It assumes that once established, "institutions can take on a life of their own and are difficult to control by those that created them" (Niemann 2006, p. 43).

The central idea of Neofunctionalism is the concept of spill-over, which can be summarised as follows:

> The central thesis of neo-functionalism is that integration within one sector will tend to beget its own impetus and spread to other sectors. The establishment of supranational institutions designed to deal with functionally specific tasks will set in motion economic, social and political processes which generate pressures toward further integration. This is the logic subsumed under the headings of "spill-over" or "the expansive logic of sector integration".
>
> (Tranholm-Mikkelsen 1991, p. 4)

While MEAs usually do not integrate additional sectors but stay in the line of regulating one specific environmental problem each, it has been established that they do integrate additional functions. While this is not the same as the far-reaching integrative processes taking place in the EU, Haas himself argued that the general logic of spill-over can be applied in a much broader sense. Indeed, the idea of Neofunctionalism was not originally a theory intended only for the European context. In his 1961 Article "International Integration: The European and the Universal Process" and in his book *Beyond the Nation-State* (1964), Haas applies his ideas on integration in a much broader context. Combining the original version of functionalism with a systems theory perspective, he develops an approach for the analysis of integration in IOs more generally and applies it to the International Labour Organisation and even broader to international welfare policy (1964) as well as to the United Nations (UN, 1961). While Haas is aware that the integrative processes at the European level cannot be reproduced in other contexts, he states that it is not clear that slightly different functional pursuits, responding to a different set of converging interests, may not also yield integration (Haas 1961, p. 389). Central here is the idea of upgrading the common interest of the parties involved, which he describes as:

> (...) the parties succeeded in redefining their conflict so as to work out a solution at a higher level, which almost invariably implies the expansion of the initial mandate or task. In terms of results, this mode of accommodation maximises what I have elsewhere called the "spill-over" effect of international decisions: policies made in carrying out an initial task and grant of power can be made real only if the task itself is expanded, as reflected in the compromises among the states interested in the task
>
> (p. 123)

As such, the transfer of the theory of Neofunctionalism to the context of MEAs seems more natural than one might expect, and its core ideas fit greatly with the understanding of MEAs employed here. Further, Haas argues that those activities have the highest spill-over potential, which involves converging interests among conflicting states. These activities require the creation of an organisation for adequate control and evoke the upgrading of common interest in the execution of highly specific programmes. The examples he mentions are clearly influenced by the zeitgeist of his time: the peaceful use of outer space, pooled space research, and UN control over extra-terrestrial bodies (Haas 1961, p. 389). From today's perspective, environmental protection is clearly also a relevant candidate for high spill-over potential in this broader sense.

While retaining its original thought, Neofunctionalism has since evolved into a much more clear-cut theory, and the concept of spill-over has evolved into a much more specific analytical concept (Niemann 2006, p. 17). Three different spill-over mechanisms have been distinguished based on their respective driving forces: functional, cultivated, and political. At its core, functional spill-over is driven by functional pressures resulting from endogenous interdependencies and embodies the original focus of the theory, cultivated spill-over is driven by the supranational institutions acting in their own interest, and political spill-over is driven by national elites and bureaucracies shifting their focus to the community level (Tranholm-Mikkelsen 1991, pp. 4–5, 6). Since political spill-over relies predominantly on the socialisation effects of the permanent large-scale bureaucracies which do not exist in MEAs, this mechanism will not be included going further. The other two, however, provide very promising insights for the study of MEAs. Two driving forces of informal MEA authority expansion based in each of them will be proposed below. This includes functional pressure based in functional spill-over and actor-driven pressure based in cultivated spill-over. Figure 3.2 summarises both pathways.

Functionally Driven Authority Expansion

The concept of functional spill-over is the structural backbone of the Neofunctionalist theory and originated in Haas' writing on functional–economic interdependencies. It starts from the recognition that some sectors within industrial economies are so interdependent that it is impossible to isolate them from the rest. It follows that the integration of one sector can only be effective if followed by the integration of other sectors. The functional integration of one task, in this regard, inevitably leads to problems which can only be solved by integrating yet more tasks (Niemann 2006, p. 17). According to Haas, earlier decisions:

> spill over into new functional contexts, involve more and more people, call for more and more interbureaucratic contact and consultation, thereby creating their own logic in favour of later decisions, meeting in a pro-community direction, the new problems which grow out of the earlier compromises.
>
> (1961, p. 371)

Lindberg (1963, p. 10) provides a wider definition of (functional) spillover:

> In its most general formulation, "spill-over" refers to a situation in which a given action, related to a specific goal, creates a situation in which the original goal can be assured only by taking further actions, which in turn create a further condition and a need for more action, and so forth. (…) The initial task and grant of power to the central institutions creates a situation or series of situations that can be dealt with only by further expanding the task and the grant of power.

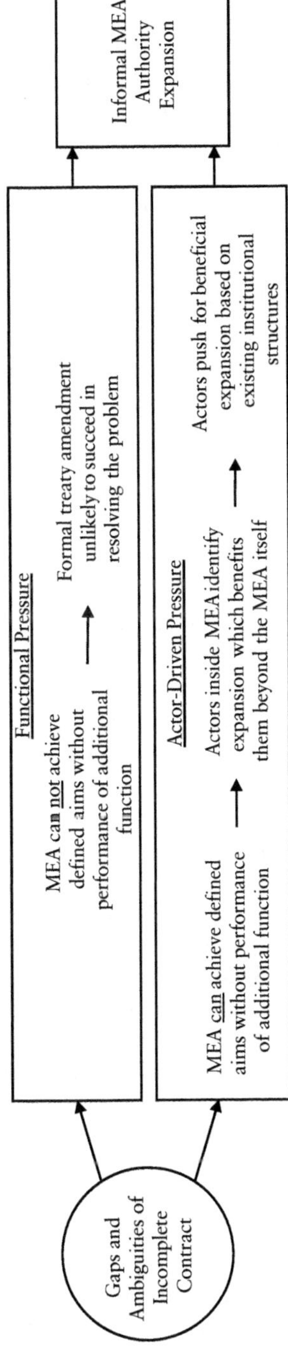

Figure 3.2 Two driving forces of informal MEA authority expansion.

From Lindberg's perspective, the dynamics of spill-over are dependent upon the fact that support for any given step in integration is the result of a convergence of goals and expectations. A lack of agreement between governments may lead to an expanded role for the central institutions; in other words, member states may delegate difficult problems. Furthermore, the activities of the central institutions may create situations that cannot be dealt with except by further central institutional development and new central policies. This mechanism assumes the continued commitment of member states to the cooperation project. However, while some states might be reluctant to deepen cooperation, the alternatives will gradually be limited, and withdrawing will become more and more difficult (as Brexit has impressively illustrated). For spill-over to happen, the will to proceed with integration does not have to be positive. Given only a general reluctance among some states, the stimulus to action can be provided by other states committed to the project or by the institutions themselves (Lindberg 1963).

Based on Lindbergs' wider definition, Niemann (2006, pp. 30–31) departs from Haas' narrow focus on economic linkage between policy areas and develops a definition of functional spill-over that encompasses all types of "endogenous-functional interdependencies", meaning all tensions and contradictions arising from within the rules of the institution, that induce policy-makers to take additional integrative steps in order to achieve their original objectives. This broader conception not only makes it possible to account for pressures to induce further integrational steps into other sectors but also for pressures to generate increased cooperation in the same field. To distinguish between these two phenomena, Niemann refers to the former as (functional) "spillover pressure" and the latter as (functional) "pressure from within" (p. 30).

The EU and MEAs differ in many ways, but the basic logic of Neofunctionalist thinking and spill-over aligns with the conceptualisation of relational authority of MEAs and authority expansion developed in the previous chapters. As outlined above, MEAs are designed to evolve into the form that best suits their specific needs over time so they can adapt to newly discovered challenges. The dynamic character of their relational authority allows them, in this context, to take on additional functions if deemed necessary by itself and its members. Since MEAs have a very narrowly defined focus and deal with environmental rather than economic issues, they are not subject to the "classic" functional spill-over leading to sector integration, as described by Haas. But they might very well be subject to pressures similar to what Niemann calls "functional pressure from within" to increase cooperation inside their specific field, resulting from the gaps and ambiguities of the incomplete contract.

More specifically, unexpected tensions might arise, which require the MEA to perform a function it is not yet mandated to perform in order to fulfil its defined purpose in a meaningful way. Functional pressure can be identified as the main driving force behind an authority expansion if the aims defined in the founding treaty MEA cannot be achieved unless action is taken to expand the cooperation project. The central question is, could the cooperation project keep running in a meaningful way if no further action was taken? If the answer is no, functional pressure is present.

In a highly dynamic issue area like environmental politics, in which much uncertainty remains, and new scientific insights regularly update our understanding of the problem, this can be expected to happen regularly, so much so that states designing the original contract usually intend for this type of mechanism to happen. The regulatory framework of MEAs generally includes a "catch-all"-phrase mandating the institution to perform such other functions as deemed necessary to reach the goals stated in the contract allowing this type of claim. If no such phrase exists, the doctrine of implied powers might provide a legal basis (see Chapter 1). While the possibility for this type of mechanism is built into the design of the organisation, it might still take unexpected turns and lead to expansions beyond what member states would otherwise be willing to give up.

H1 Functionally Driven Authority Expansion: If the MEA cannot fulfil its defined purpose without the performance of an additional function, then it can engage in performing this function without formal delegation by the member states.

To confirm functional pressure as the central driving force of an informal MEA authority expansion, it has to be shown that the MEA could not have continued to operate in a meaningful way, working toward the achievement of the aims defined in its constitutional treaty, without this informal expansion. To this end, first, it has to be established that an expansion was triggered by the occurrence of a problem, which prevented the achievement of the MEA's defined aims and thus created a functional need to take action. This is not unlikely to occur in incomplete contracts where gaps and ambiguities might only become apparent and cause trouble at later stages. While problems might be caused by a plethora of different issues, they are expected to most commonly involve topics surrounding membership and the capacity or willingness of member states to participate in the MEA. Many environmental issues entail public good problems and require participation by as many states as possible. If this functional need for action can be confirmed, second, it needs to be shown that the situation could not have been resolved through formal delegation. A central issue surrounding formal amendments is the ratification process, which can create situations in which some states might delay their ratification to avoid responsibilities, thus rendering collective measures ineffective. The argument is further strengthened if it can be shown that the expansion happened under extensive discussion of alternative options and clear discomfort of powerful states or if negotiations are driven by a sense of necessity in the light of the aims of the MEA since this supports the argument that the expansion was not pushed by political will but rather by necessity.

Actor-Driven Authority Expansion

In addition to a functional logic of integration, Haas and Lindberg also referred to the integrative impact of the European institutions themselves. These institutions have been observed to strategically attempt to cultivate relations with interest groups and national civil servants to gather their support for integrative objectives and to cultivate pressure vis-à-vis governments, particularly by upgrading common interests (e.g. through facilitating logrolling or package deals) (Niemann 2006, p. 42). Thus, the outcome of the integration process will, to some degree, be dependent on these institutions' ability and will to perform this function. Cultivated spill-over adds a voluntaristic element to an otherwise relatively deterministic theory (Tranholm-Mikkelsen 1991, p. 6)

In this context, Haas and Lindberg focused, in particular, on action taken by the Commission as an institutionalised mediator. However, the mediating services do not necessarily have to come from supranational institutions but could simply be provided by "a single person or board of experts with an autonomous range of powers" (Haas 1961, p. 368). Niemann (2006, p. 43) expanded this view in light of the changing institutional setup of the EU to incorporate institutions such as the European Parliament, the European Court of Justice, or the Council Presidency into the concept of cultivated spillover. In this way, cultivated spill-over signifies the intrinsic propensity of supranational institutions to advance the European integration process more generally. In short, cultivated spill-over describes a process in which institutional bodies consciously push for integration. In negotiations, institutional mediators focus on upgrading common interest by swapping concessions of states so that each feel that by conceding something, they have gained something else in return. Beyond that, they can also instrumentalise relationships with interest groups to gain support for further integration.

This conscious push results from the fact that the central institutions embody common interest. Thus, it is their prime role to offer solutions which upgrade this common interest by linking new proposals to the original objectives of cooperation. The Commission seems to keep saying to the Council, "if you really want the internal market, as you say you do, then you must buy this as well" (Tranholm-Mikkelsen 1991, p. 15). For the same reason, they also tend to benefit personally from progressing integration since it expands their own scope. Concerned with increasing their own powers, supranational institutions become agents of their own integration (Niemann 2006, p. 43).

Similar processes could play out inside MEAs. Since MEAs embody the common interest of protecting a specific part of the environment, actors inside the institution are in a privileged position to do so above those that want less. Since they can connect their demands to the core of the MEA's purpose, institutional structures will work in their favour. Since typical issues in the realm of environmental politics evolve around financial support and the avoidance of free riding, it seems likely that a situation might occur in which more integration might indeed come with personal benefits for some. While no direct equivalent to the Commission as central initiator and neutral mediator exists in MEAs, Niemann's (2006) opening of the concept to include all centralised institutions paved the way to transfer the mechanism. In MEAs, spill-over effects could be cultivated by different actors at the collective level, including the COP and its negotiating blocks, as well as its subsidiary bodies and expert groups.

More specifically, if functional pressure can be excluded as the main driving force, it seems that an authority expansion happened because there was an interest in expanding the cooperation on the collective level. A group of states inside the MEA could cultivate an opportunity for authority expansion, which allows them to realise a specific preference. While a single state or small group of states would most likely not be able to impose their own preferences individually, they might be able to push into a particular direction by strategically exploiting the structures of the organisation. More specifically, they can strategically employ underlying gaps and ambiguities of the incomplete contract to realise their own aims. Since the rules of the treaty are not conclusively defined, they are deployed in different ways. States can do this by connecting their demand to the achievement of the defined objectives of the MEA in the absence of actual functional pressure. Since it is the purpose of the institution to achieve this goal, it then also becomes its purpose to provide this demanded function. This allows it to be connected to existing rules and earlier decisions, which can be converted toward this new end. This can be pushed further by swapping concessions and brokering package deals to gain broader support and through the instrumentalisation of relationships with national elites, interest groups, experts, and epistemic communities.

Since the institution embodies the common interest and it is its purpose to upgrade this interest through collective solutions, opposing states tend to still benefit to some extent, even if they do not support the expansion, because they benefit from the overall persistence and efficiency of the institution. Parties recognise that they should reach for a common stand in order not to jeopardise those areas in which consensus prevails. Since compliance in MEAs is only collectively voluntary but individually mandatory, even those that question the purpose of the expansion will have to accept it because their only alternative would be to reject the institution as a whole and leave. In this way, an authority expansion can be achieved through informal ways even if there would not have been enough support for a formal amendment. Then, an informal expansion might be preferable to avoid the danger of national parliaments denying ratification.

H2: Actor-Driven Authority Expansion: If a group of states inside the MEA has an interest in an authority expansion in the absence of functional pressure, it can strategically employ the institutional structures to realise this expansion without formal delegation.

The investigation into actor-driven pressure becomes relevant once functional pressure has been excluded as the main driving force, meaning the MEA could have continued to operate in a meaningful way without the expansion. To confirm actor-driven pressure as the man driving force, it needs to be shown that the proponents of the expansion in question pursued benefits which go beyond the mere functionality of the cooperation project as such. This could include issues such as financial benefits through additional support mechanisms but also issues of membership, voting power, and control over certain decisions. Further, it has to be shown that they realised the expansion by employing the institutional structures of the MEA, where a formal amendment would have been unlikely to succeed. This can involve techniques of mobilising the aims and logics of the institution for their arguments but also the creation of alliances or the brokering of package deals inside the MEA, which would not have been possible to the same extent on the outside. Gaps and ambiguities in the existing framework make this type of arguing possible where not all details are clearly specified yet and thus might be twisted in the desired way.

Since functional pressures are anticipated to occur, and most MEAs specifically allow an expansion in cases of functional necessity, H1 encompasses the expected and legally anticipated way for authority expansions to happen and will be tested first. Only if functional pressure can be excluded as the main driving force, H2 will be investigated. If there was no functional need for an expansion, it seems likely that someone inside the MEA must have had an interest in the expansion and, therefore, strategically pushed for it. As such, authority expansion is expected to be either driven by functional pressure or – in the absence of functional pressure – driven by an actor or group of actors inside the institution that benefit from this specific expansion and, therefore, strategically cultivate pressure for it. Still, the focus lies on identifying the *main* driving force pushing an expansion. This does not mean that it will be the *only* driving force at play. It is possible that, where one is identified as the main driving force, elements of the other are at play at the same time, however, being relevant to a much lesser extent to the final outcome. Further, since the phenomenon of MEA authority expansion has so far rarely been studied, it is possible that additional driving forces or additional aspects to the driving forces proposed here will be uncovered, or that the expected driving forces turn out to operate in a different way than described.

Analysing Informal Authority Expansion

A systematic analysis of six cases of MEA authority expansion will follow to investigate these hypotheses and illustrate the phenomenon more generally. Since a phenomenon is investigated which has so far been neglected, elements of theory-testing and theory-building will be combined to see how far existing approaches can take us and where deviations from well-known paths might be needed. As such, the investigation will not stop at proving or disproving the hypotheses established above but will also remain open to the possibility of discovering unexpected factors and refining expectations along the way.

The case studies will loosely follow what George and Bennett (2005) call a "building block" study. By focussing on a specific type or subclass of a phenomenon in question, analysing this "block" fills an empty space in the overall theoretical landscape (2005, p. 76). The specific type in question is MEAs, understood as a subtype of IOs. As such, the building block provided by this analysis will contribute a new component to the broader theoretical landscape of IOs and their dynamic tendencies.

To enable cumulative development of knowledge and theory beyond the individual cases and allow for the drawing of generalisable conclusions, each case study will follow the same line of investigation. The case studies conducted here will be structured along three basic questions: First, what is the case? To begin, it needs to be established that a specific decision is a case of authority expansion as defined below and how it changes the obligations of states under the MEA. Second, can functional pressure be identified as the central driving force behind this expansion? If yes – what did this pressure look like and how did it unfold? If no – can cultivated pressure be identified and described? The observable implications defined above will provide systematic guidance to answer this question in a counterfactual approach. Third, how was the situation ultimately resolved? At this point, it will be described how the expansion in question fulfilled the functional or cultivated need that triggered its creation.

These questions will guide a systematic qualitative analysis of treaty documents, COP decisions, and related negotiation records of the COP, as well as subsidiary bodies and specialised committees, including all related submissions, statements, and additional documents. This will be supplemented by relevant scientific literature as well as economic and environmental data.

Six cases of informal authority expansion will be investigated. A case is defined as a situation in which the MEA employs, generally through a COP decision, a function (as defined in Chapter 3) that it was not previously mandated to perform, and which was not formally delegated to it through its constitutional treaty or a formal treaty amendment and subsequent ratification. The employment of this new function was not anticipated when the original contract was adopted and moved the organisation in a new direction.

This book only investigates cases in which an informal expansion took place, meaning there will be no variance in the dependent variable. While this "no variance" approach has been criticised by some, it is unavoidable here. The aim cannot be to uncover conditions under which authority expansions happen or not since current approaches in this specific field are not developed enough as of yet to set up a meaningful study in this way. Instead, the focus first needs to be on demonstrating the fact that the phenomenon does indeed exist, what it looks like, and on uncovering what makes it possible at all. Thus, the analysis will focus on typical cases in which informal authority expansion did happen, and thus, the driving force must also be present and can be identified. This is necessary first to gain a basic understanding of the phenomenon as such and its inner workings, based on which, in a later study, the conditions of its occurrence can be investigated in a more traditional way. The investigation of negative cases in which the phenomenon in question is not present and, therefore, no driving force can be identified would not be conducive to the aims of this present work.

An extensive understanding of an MEAs general history, structure, and functioning, and the underlying environmental issue is necessary for conducting the in-depth case studies according to this framework. Thus, the cases will be chosen from only two MEAs, for each of which a general introduction will be provided. The two chosen MEAs are the Vienna Convention for the Protection of the Ozone Layer and its Montreal Protocol and the United Nations Framework Convention on Climate Change with its Kyoto Protocol and Paris Agreement. The Montreal Protocol is one of the longest-existing MEAs, and it is considered the most successful to date. This makes it the logical starting point for an investigation into the nature and dynamic evolvement of MEAs. The climate MEA, on the other hand, is not only the most prominent MEA of todays' time, it has also experienced a more tumultuous evolvement and is now seen by many as having failed. Thus, a range of different contexts and conditions will be covered among the chosen cases.

The chosen cases represent the most prominent instances of authority expansion with the most far-reaching consequences for states on the international level which can be identified under each MEA to allow for an extensive, encompassing, and relevant analysis of the phenomenon. These cases are in the area of ozone protection, the creation of the Multilateral Fund, which created extensive financial obligations for members of the protocol, the creation of the compliance committee, which introduced the managerial approach to environmental politics, and the peculiar case of the definition of developing countries which turned out to be more difficult to solve than one might expect. In the area of climate protection, these are the creation of the unusually strict compliance mechanism is analysed in contrast to the Montreal approach as well as the creation of the Green Climate Fund, which represents an entirely new organisation created by an MEA, and the Adaptation Fund which was a surprising win by developing countries.

References

Brunnée, Jutta (2004): The United States and International Environmental Law. Living with an Elephant. *European Journal of International Law* 15 (4), pp. 617–649. DOI: 10.1093/ejil/15.4.617

Cooper, Scott; Hawkins, Darren; Jacoby, Wade; Nielson, Daniel (2008): Yielding Sovereignty to International Institutions. Bringing System Structure Back In. *International Studies Review* 10 (3), pp. 501–524. DOI: 10.1111/j.1468-2486.2008.00802.x

Gehring, Thomas (2002): *Die Europäische Union als komplexe internationale Organisation. Wie durch Kommunikation und Entscheidung soziale Ordnung entsteht*. Baden-Baden: Nomos.

Gehring, Thomas (2008): Treaty-Making and Treaty Evolution. In Daniel Bodansky, Jutta Brunnée, Ellen Hey (Eds.): *The Oxford Handbook of International Environmental Law*. Oxford: Oxford University Press.

George, Alexander L.; Bennett, Andrew (2005): *Case Studies and Theory Development in the Social Sciences*. Cambridge, MA: The MIT Press.

Haas, Ernst B. (1958): *The Uniting of Europe. Political, Social, and Economical Forces: 1950-1957*. Notre Dame, IN: University of Notre Dame Press.

Haas, Ernst B. (1961): International Integration. The European and the Universal Process. *International Organization* 15 (3), pp. 366–392. DOI: 10.1017/S0020818300002198

Haas, Ernst B. (1964): *Beyond the Nation State*. Stanford, CA: Stanford University Press.

Hawkins, Darren G.; Lake, David A.; Nielson, Daniel L.; Tierney, Michael J. (2006): *Delegation and Agency in International Organizations*. Cambridge: Cambridge University Press.

Heldt, Eugénia; Schmidtke, Henning (2017): Measuring the Empowerment of International Organizations. The Evolution of Financial and Staff Capabilities. *Global policy* 8 (5), pp. 51–61. DOI: 10.1111/1758-5899.12449

Koch, Martin (2009): Autonomization of IGOs. *International Political Sociology* 3 (4), pp. 431–448. DOI: 10.1111/j.1749-5687.2009.00085.x

Lake, David A. (2010): Rightful Rules. Authority, Order, and the Foundations of Global Governance. *International Studies Quarterly* 54 (3), pp. 587–613. DOI: 10.1111/j.1468-2478.2010.00601.x

Lindberg, Leon N. (1963): *The Political Dynamics of European Economic Integration*. Stanford, CA: Stanford University Press.

Lindberg, Leon N.; Scheingold, Stuart A. (1970): *Europe's Would-Be Polity. Patterns of Change in the European Community*. Englewood Cliffs, NJ: Prentice-Hall.

Luhmann, Niklas (1987): *Soziale Systeme. Grundriss einer allgemeinen Theorie*. Frankfurt am Main: Suhrkamp.

Luhmann, Niklas (2021): *Die Gesellschaft der Gesellschaften*. Frankfurt am Main: Suhrkamp.

Mahoney, James; Thelen, Kathleen (2009): *Explaining Institutional Change*. Cambridge: Cambridge University Press.

Moravcsik, Andrew (1993): Preferences and Power in the European Community. A Liberal Intergovernmentalist Approach. *Journal of Common Market Studies* 31 (4), pp. 473–524. DOI: 10.1111/j.1468-5965.1993.tb00477.x

Moravcsik, Andrew (1998): The Choice for Europe. Social Purpose and State Power from Messina to Maastricht. Ithaca, NY: Cornell University Press.

Niemann, Arne (2006): *Explaining Decisions in the European Union*. Cambridge: Cambridge University Press.

Pierson, Paul (2000): The Limits of Design. Explaining Institutional Origins and Change. *Governance* 13 (4), pp. 475–499. DOI: 10.1111/0952-1895.00142

Rocco, Philip; Thurston, Chloe (2014): From Metaphors to Measures. Observable Indicators of Gradual Institutional Change. *Journal of Public Policy* 34 (1), pp. 35–62. DOI: 10.1017/S0143814X13000305

Thelen, Kathleen (2003): How Institutions Evolve. Insights from Comparative Historical Analysis. In James Mahoney, Dietrich Rueschemeyer (Eds.): *Comparative Historical Analysis in the Social Sciences*: Cambridge: Cambridge University Press, pp. 208–240.

Thelen, Kathleen Ann; Streeck, Wolfgang (Eds.) (2005): *Beyond Continuity. Institutional Change in Advanced Political Economies*. Oxford: Oxford University Press.

Tranholm-Mikkelsen, Jeppe (1991): Neo-Functionalism. Obstinate or Obsolete? A Reappraisal in the Light of the New Dynamism of the EC. *Millennium* 20 (1), pp. 1–22. DOI: 10.1177/03058298910200010201

Werksman, Jacob (2014): *Greening International Institutions*: London: Routledge.

Wiersema, Annecoos (2009): The New International Law-Makers? Conferences of the Parties to Multilateral Environmental Agreements. In *Michigan Journal of International Law* 31 (1), pp. 231–287.

Zürn, Michael; Binder, Martin; Ecker-Ehrhardt, Matthias (2012): International Authority and Its Politicization. *International Theory* 4 (1), pp. 69–106. DOI: 10.1017/S1752971912000012

4 The Montreal Protocol on Substances that Deplete the Ozone Layer

The ozone layer is a region in our planet's stratosphere, 15–35 km above the surface, which contains high concentrations of the ozone (O3) molecule. It is essential for life on planet Earth as it absorbs most of the harmful ultraviolet (UV) radiation emitted by the sun. Thus, it came as a shock when it was discovered in the 1970s that the emission of chlorofluorocarbons (CFCs) into the stratosphere was causing dangerous damage. At the time, CFCs were rapidly proliferating compounds with wide applications in thousands of products, such as refrigeration, air conditioning, aerosol sprays, solvents, transportation, plastics, insulation, pharmaceuticals, computers, electronics, and firefighting. Soon, however, negotiations on a Multilateral Environmental Agreement (MEA) which would phase down their use would be adopted. Today, the Montreal Protocol on Substances that Deplete the Ozone Layer, signed on September 16, 1987, is widely regarded as the most successful MEA to date (Rajamani 2012, p. 608). 198 states have ratified the Montreal Protocol, 99 per cent of the controlled ozone-depleting substances have been phased out, and there is scientific evidence that the ozone layer is healing (UNEP 2020). Since its adoption, its institutional structures have extensively expanded over time and set many precedents for future MEAs and their design. As such, the Montreal Protocol represents the ideal starting point for an analysis of MEA authority expansions. After a short introduction to the issue of ozone depletion and the emergence of the Montreal Protocol, three cases of authority expansion of the protocol will be investigated in detail to discover their driving forces.

Explained in simple terms, ozone (O3) is an unstable molecule composed of three, rather than the customary two, oxygen atoms. When an ozone molecule absorbs UV radiation, it splits into an oxygen molecule (O2) and a separate oxygen atom (O). Later, the two components reform the ozone molecule (O3) until they are broken up again. In this way, ozone is continuously created, destroyed, and recreated. When undisturbed, a consistent balance of ozone in the stratosphere is maintained (UNEP 2020; Chapman 1930).

However, in 1974, two chemists at the University of California, Mario Molina and Sherwood Rowland, discovered that human activity had

DOI: 10.4324/9781003597681-5

begun to interfere with this process. A family of widely used anthropogenic chemicals, the CFCs, had been added to the natural environment through human activity in steadily increasing amounts over the preceding decades. Molina and Rowland discovered that, unlike most other gases, CFCs are not chemically broken down or rained out quickly in the lower atmosphere but rather, because of their exceptional stability, persist and migrate slowly up to the stratosphere, where they are eventually broken down by solar radiation and release large quantities of chlorine. This chlorine interferes with the natural balance of ozone creation and recreation so that it is destroyed more quickly than it can be reproduced, leading to a depletion of the ozone layer. Simply speaking, the chlorine (Cl) takes one of the three oxygen atoms from an ozone (O3) molecule and moves it to a single oxygen atom (O). In this way, three molecules of the more stable O2 are produced from two molecules of ozone, regenerating the chlorine atom at the end of the process, ready to restart the process (Molina and Rowland 1974; Benedick 1998, pp. 10–16; Parson 2003, pp. 17–22). Catalysed by chlorine in this way, ozone is destroyed more quickly than it can be naturally recreated. One chlorine atom can destroy over 100,000 ozone molecules before it is removed from the stratosphere (EPA 2018).

Importantly, it does not matter where in the world and by which means the CFC molecules are released. As a result of their long residence time in the atmosphere, they are transported globally and have global effects (Rowlands 2008, p. 322).

Although uncertainty about these findings was high at first, it soon became clear that if it was real, the consequences would be devastating. Projections indicated that an 1 per cent depletion of the ozone layer could be reached as early as the year 2000, and 1 per cent would already make a big difference. A link between exposure to UV-B radiation and skin cancer had already been established. It was calculated that 1 per cent ozone depletion would lead to an increase in non-melanoma skin cancer incidence of 4.8 per cent and an increase in melanoma incidence of 1–2 per cent. Based on the middle scenario of projected future ozone depletion in the absence of regulation, this would mean approximately 153 million additional cases and 3 million additional deaths of nonmelanoma skin cancer and an additional 782,100 cases and 187,000 additional deaths of melanoma for the US population alive at the time and born by 2075. UV-B radiation had also been linked to eye damage. For the same time frame and population, an 1 per cent depletion would mean 18.2 million additional cases of eye cataracts. Limited studies also suggested that UV-B radiation induces suppression of the immune system. In addition to these health issues, major damage to agriculture and fisheries and accelerated weathering of outdoor plastics were also suspected (EPA 1987). Because of their long lifetimes, as much as nine-tenths of all CFCs ever emitted were still present in the atmosphere at the time. And, millions of additional tonnes of CFCs were already on their way up. Even if CFC emissions were to level off or decline immediately, chlorine would continue to accumulate in the stratosphere for decades (Benedick 1998, p. 11).

The discovery came as an environmental as well as an economic bombshell. There had been no prior suspicion that CFCs were harmful in any way. CFCs had been thoroughly tested and found to be safe by customary standards. The possibility that dangers could emerge many miles above the Earth's surface had simply never been considered. Indeed, following their invention in the 1930s, CFCs had seemed an ideal chemical. CFCs are stable, non-flammable, non-toxic, and non-corrosive – qualities that make them extremely useful in many industries. They often replaced other chemicals, such as ammonia, in refrigerators, the dangers of which were widely known. They are energy-efficient coolants in refrigerators and air conditioners, as well as effective propellants in spray containers for cosmetics, household products, pharmaceuticals, and cleaners. They were also used as insulators and are standard ingredients in the manufacture of a wide range of rigid and flexible plastic foam materials. Their non-reactive properties make them convenient solvents for cleaning microchips and telecommunications equipment and for use in a number of other industrial applications. They are also inexpensive to produce. And thus, their production had soared (Benedick 1998, p. 11).

Scenarios were disturbing enough to call states into action. Since CFCs remain in the atmosphere for about 100 years, stabilising the ozone layer would require not merely a freeze but drastic cuts in CFC production and consumption. While the industry was reluctant to let them go, it soon became clear that while the relevant products were inexpensive and extremely useful, they were not particularly profitable and less harmful alternatives were soon discovered. Thus, abandoning CFCs to protect the ozone layer was clearly inconvenient, but ultimately, the costs seemed relatively low compared to the dangers of continued depletion (Parson 2003, pp. 156–159).

In the late 1970s, many industrialised countries began to unilaterally restrict CFC consumption and production, albeit with different degrees of seriousness. However, it became clear early on that geography did not matter much for ozone destruction. As a problem affecting the entire world, any lasting solution had to take place in a global context (Benedick 1998, p. 40). Developing countries, however, were more reluctant to engage in the reduction effort. At the time, they only consumed an extremely small percentage of the substances in question. Many of them, however, like China and India, had the potential for rapidly growing domestic markets (Gehring 1994, p. 271). Many of the substances in question were intimately linked to development and, as they would later argue, would be essential to raising their living standards (Rosencranz and Milligan 1990, p. 313). Indeed, this issue would influence not just the original rules of the Montreal Protocol but also dominate subsequent discussions of expansions.

In January 1982, the UN Environment Programme (UNEP) convened representatives of 24 countries in Stockholm to launch the Ad Hoc Working Group of Legal and Technical Experts for the Preparation of a Global

Framework Convention for the Protection of the Ozone Layer. Soon after, on March 22, 1985, the Vienna Convention for the Protection of the Ozone Layer was signed by 20 (mostly industrialised) nations and the European Community (EC). In its preamble, states set their intention to "protect human health and the environment against adverse effects resulting from modifications of the ozone layer". The convention creates a general obligation for states to take "appropriate measures" to achieve this aim (Article 2.1). Although no specific measures are defined, several areas of action are mentioned (Article 2.2), including: (a) cooperation in research and information exchange, (b) the adoption and harmonisation of policies to control, limit, reduce, or prevent human activities which have or are likely to have adverse effects on the ozone layer, (c) co-operation in the formulation of agreed measures, procedures, and standards for the implementation of the convention, and (d) co-operation with competent international bodies to implement the convention and its protocols effectively. The Vienna Convention establishes a Conference of the Parties (COP), which meets at regular intervals to keep under continuous review the implementation of the convention, review scientific information, promote harmonisation of policies and strategies, make recommendations, adopt programmes for research and technology transfer, and consider and adopt protocols as well as amendments and annexes to the convention and its protocols, establish subsidiary bodies, and seek the services of competent international bodies (Article 6). The COP is also mandated to "consider and undertake any additional action that may be required for the achievement of the purposes of this Convention" (Article 6.4 (k)). Further, the convention includes an elaborate mechanism to settle disputes among parties (Article 11). Symbolising the reluctance of many participating governments at this stage, however, was the fact that nowhere did the Vienna Convention specifically identify any chemical as an ozone-depleting substance (Benedick 1998, p. 45).

Specific obligations would, however, soon be added in the form of a protocol. On September 16, 1987, representatives of 24 nations, plus the Commission of the EC, signed the Montreal Protocol on Substances that Deplete the Ozone Layer (Benedick 1998, p. 98). Building on the aims and obligations of the convention, parties to the protocol commit themselves in the preamble to:

> protect the ozone layer by taking precautionary measures to control equitably total global emissions of substances that deplete it, with the ultimate objective of their elimination on the basis of developments in scientific knowledge, taking into account technical and economic considerations.

The primary obligation of all parties to the protocol is the timely phase-out of specified ozone-depleting substances in accordance with specified control schedules (Article 2). Originally, the Montreal Protocol contained specific obligations for the reductions of five CFCs and three halons, which were suspected of causing similar processes to CFCs in the stratosphere. The restrictions apply to a basket of chemicals, each weighted by its "ozone-depleting potential", to provide flexibility and lower the overall costs of ozone protection. To handle a concern that states might have an incentive to stay outside of the treaty and fill the supply gap resulting from these control measures, thus undermining all efforts, the protocol prohibits imports from, and exports to, non-parties (except those non-parties that are otherwise in full compliance with the protocol's reduction schedules) (Article 4) (Rowlands 2008, p. 326).

To address the concerns of developing countries that adhering to the strict control measures could compromise their development, the Montreal Protocol includes special provisions for their participation in its Article 5. Specifically, paragraph 1 grants a 10-year grace period for compliance with the control measures for those that are developing countries and "whose annual calculated level of consumption of the controlled substances is less than 0.3 kilograms per capita". Those parties to the protocol falling under this condition are referred to as "Article 5 parties". Further, paragraph 3 aims at the facilitation of bilateral or multilateral financial assistance, although no specific structures for financial support were established at this point.

The Montreal Protocol establishes its own membership body, the Meeting of the Parties (MOP), separate from the COP of the convention (Article 11). The MOP meets at regular intervals to review the implementation of the protocol, establish guidelines for reporting, and review requests for technical assistance and reports. It also has a particularly strong influence on the control measures, including the adjustment of existing control measures and the addition or removal of substances to the list in accordance with the conventions' amendment procedure. While amendments require ratification of the individual parties, adjustments only require consensus of the MOP. This is one of only a few provisions which particularly require consensus. On all other matters of substance, the MOP takes decisions by a two-thirds majority, meaning a significant loss of influence for individual parties. Finally, the MOP is also mandated to "consider and undertake any additional action that may be required for the achievement of the purposes" of the protocol (Article 11.4 (j)).

Once the Montreal Protocol entered into force, ongoing assessments – both scientific and economic – played a critical role. The specific architecture of the protocol allowed new information to be fed into the existing structure so that adjustments could be discussed and incorporated quickly (Rowlands 2008, p. 323). It was deliberately designed to be reopened and adjusted as needed on the basis of the periodically scheduled scientific, economic, environmental, and technological assessments. This flexibility was a major innovation at the time (Benedick 1998, p. 7). In this way, regulations were continuously tightened, and many of the skeletal provisions of the original protocol were fleshed out in the years that followed – including many instances of authority expansion.

Several formal amendments expanded the Montreal Protocol in a formal way. The most relevant for the operations of the protocol is certainly the London Amendment, which was adopted soon after the protocol adoption in 1990 and filled many of the gaps that were left open in the original contract. It introduced a significant tightening of the control measures, adding a significant number of new substances, and it formally established the financial mechanism with its Multilateral Fund, which would provide financial assistance to Article 5 parties. Later amendments further expanded control measures to include additional substances and alter the details of the protocol's rules. Particularly interesting is also the Kigali amendment, which added hydrofluorocarbons (HFCs) to the list of controlled substances. HCFs are not, themselves, ozone-depleting substances, but they are very potent greenhouse gases that contribute to climate change. Since they were introduced to replace the ozone-damaging CFCs in the earlier days of ozone protection (UNEP 2023), their phase-down was addressed not under the climate MEA but by the Montreal Protocol.

The Montreal Protocol has also gone through significant informal changes over the decades. Particularly in its earlier days when the skeletal rules of the contract were in the process of gaining substance, the MEA faced far-reaching challenges, the solutions to which carried it into new and unexpected directions. In this process, many striking instances of informal authority expansion can be observed. While many are technical in nature and concern the details of how specific control measures are to be implemented, three cases can be identified which are particularly far-reaching and would not only substantially change the obligations of member states but also influence how later MEAs would approach similar issues. These three cases will be analysed in detail in the following chapters to uncover their driving forces and draw conclusions on why informal expansions become possible.

First, the creation of the Interim Multilateral Fund introduced financial obligations for industrialised country parties and promised developing country parties support if they accepted the strict and further tightening obligations of the protocol. The section "The Creation of the Interim Multilateral Fund" argues that the creation of this fund was driven by functional pressure resulting from a regulatory gap in financial implementation assistance. While the participation of as many states as possible globally was necessary to make a meaningful approach to solving the problem of ozone depletion possible, developing countries would not be capable (or at least not willing) to participate without support. They argued that they were not the creators of the problem and, thus, should not have to compromise the rising of their living standards to solve it. Since the need for their participation was strong, developing countries were put in an unusually strong position. As they would not accept an arrangement under the weighted voting structures of the World Bank, the creation of an entirely new fund under the authority of the protocol with its equal representation became necessary.

Second, the decision to define developing countries on a case-by-case basis by MOP decision gave the MEA the authority to decide which parties would be able to benefit from the special provisions of Article 5 and support from the Multilateral Fund and which would be obligated to pay for it instead. While clearly differentiating between obligations and opportunities for developing and industrialised country parties, it had not been clearly established how these two groups would be distinguished. The section "Defining Developing Countries" argues that the reason for handling the situation in this unexpected way was a functional spill-over resulting from this ambiguity and the London Amendment, which much increased the significance of this division through the introduction of additional obligations and support opportunities. Existing lists held by other organisations were considered but ultimately not applied. It was feared that a number of states could be dissatisfied with their classification or incapable of adhering to the corresponding responsibilities and thus withhold their participation.

Third, the creation of the Montreal Protocols' compliance mechanism significantly expanded the MEAs' authority to handle situations of non-compliance and established an entirely new system with its own subsidiary committee. This committee is the first to follow a facilitative, managerial approach instead of the more common enforcement approach. With this, the Montreal Protocol did not only enhance its own authority but also created a precedent for how MEAs would approach compliance issues in the future. The section "Compliance Management" argues that the creation of this new mechanism became necessary because of gaps in the existing dispute settlement procedure as well as the more common enforcement approach. They both would be insufficient to address the collective nature of the problem of ozone depletion or many environmental issues more broadly. Where non-compliance hurts not another party but the collective environment, no individually injured party could bring a claim to a more traditional mechanism. Further, when addressing environmental issues, prevention of harm is more important than its punishment. For an MEA to function, no state can be discouraged from participation by strict sanctions if they encounter difficulties, but they rather need to be encouraged to stay committed through the provision of additional support.

The Creation of the Interim Multilateral Fund

The creation of the Interim Multilateral Fund of the Montreal Protocol is one of the most impressive cases of authority expansion of any MEA to date. Although later formalised through a treaty amendment, the fund began its operation based solely on a MOP decision which set up the structures and commitments of this multimillion-dollar mechanism. It created new and unforeseen financial obligations for industrialised country parties and extensive support opportunities for developing country parties. But, it would also have implications far beyond the ozone layer. Negotiators were aware at the time that they were establishing a blueprint for mechanisms in other environmental treaties, in particular the climate agreement, which gave negotiations a special taste of magnitude.

The creation of the Interim Multilateral Fund extensively broadens the MOPs' authority in the area of implementation assistance by giving it the capability to collect financial contributions from industrialised states and disburse them to developing countries through a specialised subsidiary committee from a collective point of view. This development is particularly puzzling since the original version of the Montreal Protocol did not include any type of financial mechanism at all. Even more impressively, it came to life despite the strict opposition of many of the most powerful industrialised countries.

Why was this new fund created under the authority of the MOP of the Montreal Protocol? This chapter argues that its creation was driven by strong functional pressure resulting from a mismatch between strict obligations (which were soon set to tighten even further) and the lack of commitment to financial assistance for building the necessary capacities in the developing world. Many developing countries put their participation on hold until they could trust that they would receive support. However, their participation was essential for the MEA to function effectively. Many developing countries, in particular India and China, had immense potential to grow and expand their emissions. Thus, they needed to be included, or all efforts under the protocol would be undermined. This strong need puts developing country parties in a favourable position. As such, the discussion not only became about the provision of support as such, but also about how this support would be managed. Since developing countries would not accept a structure administrated by the donor-driven World Bank, creating a new fund under the authority of the MOP became the only viable solution, even though most industrialised country parties had strictly opposed this option. Further, establishing this new fund by formal amendment alone was not an option. An amendment would only become binding on those that ratified it. If large donors delayed their ratification, others would end up having to finance a much larger share of the overall cost. Thus, an interim arrangement that would become immediately binding on all parties became necessary. And, this could only be provided by a MOP decision.

The Case: Expanding Assistance

The original version of the protocol did not foresee any type of centralised funding mechanism, particularly not one as extensive as that which was created. The only part of the treaty which explicitly mentions financial support for developing countries is Article 5 (3):

> The Parties undertake to facilitate bilaterally or multilaterally the provision of subsidies, aid, credits, guarantees, or insurance programmes to

> Parties that are developing countries for the use of alternative technology and for substitute products.

Despite this acknowledgement of the special needs of developing countries, this paragraph alone does not provide any assurance that they would receive any significant monetary or technological assistance (Patlis 1992, p. 193).

At their first meeting in Helsinki in 1989, however, the Parties agreed:

> to recognise the urgent need to establish international financial and other mechanisms (...) to enable developing countries to meet the requirements of the present and a future strengthened Protocol
>
> (Decision I/13 (1))

To develop the modalities of such a mechanism (among other refinements of the protocol), the MOP established an open-ended Working Group which soon began to explore options "which do not exclude the possibility of an international Fund" (Decision I/13 (2)). Negotiators were aware at the time that "the mechanisms designed for the Montreal Protocol would likely provide the blueprint for a mechanism to control greenhouse gas emission and for adaptation to climate change" (UNEP/OzL.Pro.WG.I(1)/3). There is no doubt that this fact was present in the minds of negotiators on both sides. They were essentially not only negotiating the Montreal fund but also all environmental funds that were to come.

At its second meeting in 1990 in London, the MOP adopted two instruments which would implement the results of these negotiations: the London Amendment, which would formally establish the Multilateral Fund, and Decision II/8, which would put into place arrangements for the interim time until the amendment would be ratified and enter into force:

> The Second Meeting of the Parties decided in Decision II/8 To establish for the three-year period from 1 January/1991 to 31/December/1993 or until such time as the Financial Mechanism is established, an Interim Financial Mechanism according to the following:
>
> The Interim Financial Mechanism is established for the purposes of providing financial and technical cooperation, including the transfer of technologies, to Parties operating under paragraph 1 of Article 5 of the Montreal Protocol to enable their compliance with the control measures set out in Articles 2A to 2E of the Protocol. (...)
>
> The Mechanism established under paragraph 1 shall include a Multilateral Fund. (...)

The Interim Fund started to operate immediately, and obligations for states to pay in became immediately binding. Although officially only set up for the interim period, the Interim Fund already included all relevant rules and structures necessary for the operation of the actual fund. While these rules and structures would not automatically apply to the regular fund, all legal instruments were drafted in a way that would allow their immediate transfer without further modification. Changes or even a re-negotiation were not expected unless a new consensus emerged between contributing and benefiting countries (Gehring 1994, p. 303). Indeed, when the Multilateral Fund was formally established at MOP 4 in 1992 in Copenhagen, the terms of the Interim Fund were carried over without changes. Even though some states tried to postpone or even stop this step and redirect the Montreal implementation assistance through other channels, they did not succeed in pushing for changes in the already existing structures (UNEP/OzL.Po/WG.1/7/4 paras. 108, 117, 118). As such, the fund was essentially established through a decision made by the MEA itself, and the formal amendment merely served as an ex-post legalisation of a particularly adventurous informal authority expansion.

The Multilateral Fund operates under the authority of the MOP and is governed by an Executive Committee consisting of 14 members selected separately on the basis of a balanced representation of industrialised and developing country parties. The Executive Committee develops and monitors the implementation of specific operational policies, guidelines, and administrative arrangements, including the disbursement of resources, for the purpose of achieving the objectives of the Multilateral Fund. It takes decisions by consensus whenever possible and can resort to a double majority of two-thirds of all parties present and voting, representing a majority of the donor countries and a majority of recipient countries, ensuring equal power for both groups. The Executive Committee cooperates with the World Bank, the UNEP and the United Nations Development Programme (UNDP). UNEP performs functions in the field of research and data gathering and acts as a clearing house, UNDP elaborates feasibility and pre-investment studies and provides technical assistance (Gehring 1994, pp. 301–302; Benedick 1998, p. 161). The World Bank was designated to be the Administrator of the programme to finance the agreed incremental costs of developing country parties, achieving a relatively strong position while still remaining under the authority of the Executive Committee (Second Meeting of the Parties, Annex IV – Appendix IV, D., para. 15).

The Multilateral Fund provides financial assistance to all member state parties operating under Article 5 (1) of the protocol, meaning all those that are classified under the protocol as developing countries and their annual calculated level of consumption of controlled substances is less than 0.3 kg per capita on the date of entry into force of the protocol for it, or any time thereafter within 10 years of the date of entry into force of the protocol. These are also the parties that are eligible for grace periods when implementing control measures. Currently, 147 of the 197 parties to the Montreal Protocol meet these criteria. Assistance is provided to meet, on a grant or concessional basis, the agreed incremental cost to enable these parties to comply with their obligations under the protocol. Contributions to the fund are additional to previously existing financial transfers and assessed on the UN scale of assessment. The initial capital of the fund was US$160 million over 3 years from 1991 to 1993, to be increased up to US$240 million upon accession of additional developing countries, in particular India and China (MOP2 Annex IV – Appendix IV: Terms of reference for the Interim Multilateral Fund). As of December 2022, over US$ 5.2 billion have been pledged to the fund (Multilateral Fund 2022).

The creation of this fund significantly enhanced the authority of the Montreal Protocol. Where originally, industrialised country parties had no specific obligations to provide any financial aid in the area of ozone protection, they now have to pay a specific and not insignificant amount, which rose with more developing countries joining the protocol. Further, the authority over how this money is used is not in their hands but in the hands of a specialised Executive Committee, a subsidiary body with equal power of developed and developing countries. In order to clarify the nature and legal status of the fund as a treaty body under international law, the 1994 Sixth MOP adopted Decision VI/16, which provides that the fund enjoys "legal capacity as is necessary for the exercise of its functions and the protection of its interests… to acquire and dispose of movable property and to institute legal proceedings in defence of its interest". With legal capacity of its own, the fund has significant independence of the parties individually. Developing countries, on the other hand, benefit hugely from this development. They now receive this money to finance the achievement of their ozone obligations, and their representation in the Executive Committee puts them on equal standing with industrialised nations in the control they can exert over aid flows. They did, however, have to give something in return. The second important component of the London Amendment was the strengthened control measures, which were the main interest of the industrialised country parties at this point in time. The London Amendment added phase-out schedules for several substances which were not controlled before. It could only be accepted or rejected as a whole, essentially making it a package deal of financial aid in exchange for stronger control measures. Thus, developing countries had to take on not just the original control measures but these now tightened obligations, albeit with grace periods.

Functional Pressure: The Need to Pay the Price of Developing Country Participation

The Montreal Protocol had started off as largely a club of industrialised countries. At the time of the first MOP in 1989, it only counted 37 parties, and only 9 of them were developing countries, most of them with low relevance for the market of ozone-depleting substances (Gehring 1994, p. 270). One of the premises of the Montreal Protocol had been that developing countries would be encouraged to join if they could count on a reasonable expansion in the use of controlled substances in a 10-year transitional period, after which they would move to newly developed technologies and follow the original reduction schedule. Negotiations over Article 5 reflected that their preoccupation was primarily to maintain maximum usage of CFCs for the longest possible time (Benedick 1998, p. 148). However, most developing countries remained reluctant to join.

Many of the controlled substances are intimately linked to development. They were used in the products associated with modernisation and social status: refrigerators, cold storage, air conditioners, aerosol sprays, foam insulation, foam cushioning, food packaging, medical supplies, computer chips, dry cleaning, and fire-extinguishing systems. Therefore, developing countries viewed the Montreal Protocol as an impediment to their drive for modernisation and industrialisation and argued that their economies would not be able to bear the high costs of switching to less ozone-damaging technologies (Rosencranz and Milligan 1990, p. 313). Governments of developing nations, China and India in particular, argued that products made with or containing CFCs, especially for food preservation and air conditioning, were essential to raising their living standards. Delegates stated that it was unacceptable for developing countries to either have to forgo these necessities or to pay more for substitutes and thereby further enrich the chemical industries that had created the ozone problem in the first place. Developing countries further stressed that the problem was not of their making. In effect, the use of these chemicals had for decades contributed to the well-being of the industrialised countries but, at the same time, had built up a threat to the entire planet in the form of long-lasting CFCs in the stratosphere (Benedick 1998, p. 149).

Further, since the adoption of the protocol, the scientific community had established a new consensus: The threat of ozone depletion was far more significant than originally predicted and much stricter cuts in ozone-depleting substances would be needed. This moved the issue of developing country participation to centre stage and would allow them to claim a major role in the revision of the protocol. With industrialised countries now on a fast track toward phase-out rather than the original 50 per cent reduction, the grace period became almost irrelevant. Instead, it would now be in the interest of developing countries not to linger too long with CFCs but rather move as rapidly as possible to new technologies. And, this move would be expensive (Benedick 1998, p. 148).

It remains unclear if the cost would indeed have been unmanageable for developing countries. No estimates of the costs they would face existed at the time, and developing countries put significant pressure on industrialised nations to take action before more reliable data would be available (UNEP/OzL.Pro.WG.I(1)/3, paragraph 16). Early estimates by the Indian government amounted to US$ 5 billion. However, it is generally agreed that this was a gross overestimate – likely for bargaining purposes. Preliminary estimates for the USA, at the time the world's largest CFC user and producer, indicated costs of US$ 2.7 billion, and later estimates for India amounted to US$ 1.2 billion. The main problems were not the cost of redesign and retooling, which can be attributed to the cost of doing business. Manufacturers in developing countries claimed that they would not be able to afford the royalty and licensing fees associated with substitute technologies (Rosencranz and Milligan 1990, p. 313). What would matter most for negotiations, however, was not the actual cost but developing countries' willingness to bear them.

Developing countries eventually announced that they would make their participation conditional on a promise of financial assistance. This came as a shock to the rest of the world (Benedick 1998, p. 148). If industrialised country parties refused to hear these concerns, it was very likely that developing countries would not have joined the Montreal Protocol. This would have seriously undermined the success of the global effort to save the ozone layer. Developing countries only consumed an extremely small percentage of the substances in question at the time. Many of them, however, like China and India, had the potential for rapidly growing domestic markets (Gehring 1994, p. 271). Population growth and their eagerness to modernise were creating a growing demand for appliances and consumer products produced with ozone-depleting substances which "could undermine the Montreal Protocol", and it was feared that "If India and China achieve their modernisation and electrification goals for the next decade, all CFC abatement efforts by the industrialised nations would be nullified" (Rosencranz and Milligan 1990, p. 312). It was estimated that "New CFC emissions from the developing nations could completely offset the major investments by firms in the industrialised nations to reduce CFC emissions" (Phillips 1990, p. 318). Detailed models prepared for the US Environmental Protection Agency (EPA) predicted that "(...) consumption of CFC-intensive products is likely to increase at a faster rate than that of per capita GNP in the developing countries for the rest of the century" (Kohler et al. 1987, p. 65). In addition to India and China as the main relevant developing countries, Brazil, Saudi Arabia, South Korea, Indonesia, Nigeria, South Africa, and Mexico were identified as those most relevant to consumption growth and Turkey, Argentina, Venezuela, Algeria as smaller countries with still important levels of consumption and Iran, Iraq, and South Africa as state with the potential to be significant consumers of CFC intensive goods but with a high level of uncertainty for projections resulting from revolution, war, and civil strife (Kohler et al. 1987, p. 6). Of those, only Mexico and Nigeria had joined the Montreal Protocol by 1989.

While trade in controlled substances and relevant products would be prohibited under the protocol, there was a possibility that developing countries would be able to meet their growing demands autonomously by producing domestically and trading amongst themselves, establishing a viable exchange outside of the protocol (Benedick 1998, p. 151). In fact, India had already begun to export refrigerators to the Soviet Union and the Middle East. They were also transferring technology for the establishment of a CFC production plant in Iraq. And, it was likely that they would be able to produce enough of the substances in question to fulfil their own domestic demands without imports from those states that became party to the protocol. Their demand in 1990 was only 5000 tonnes, while their factories were capable of producing 18,000 tonnes of CFCs annually, and a new CFC plant had just been established. Official government approval for the expansion of production was given for domestic use. With the trend towards increased use of products containing CFCs and no plan to join the Montreal Protocol, annual domestic production of CFCs was expected to increase to 25,000 tonnes in the near future. At the same time, China was planning significant expansions of CFC consumption as well, in particular in CFC-using refrigerators. The estimated increase in CFC use would have made China the world's largest CFC producer and consumer, entirely offsetting the effects of the USA as the world's largest producer at the time cutting its CFC use in half (Rosencranz and Milligan 1990, pp. 313–315).

As such, the missing promise of financial implementation assistance in the original Montreal Protocol was about to become the trigger for its demise if not resolved quickly. Since these countries were incapable of adhering to the common standards (or at least unwilling to risk compromising their development to this extent), but their participation was essential to the success of the protocol, there was a strong functional need to expand implementation assistance. Something needed to be done to pay the price for their participation. Decisions taken at MOP1 reflect the awareness of this necessity among industrialised states. According to Decision I/5 (1), the MOP:

> recognises the urgent need to establish international financial and other mechanisms to implement Article 5 (…) to enable developing countries to meet the requirements of the present and future strengthened Protocol.

This was supported by the Economic Assessment Panel of the Montreal Protocol:

> Development assistance will be required in most cases. Funding is needed for the transfer of technology during the transition period because currently available resources are already strained as a result of the world debt problem and the dire economic situation of many countries. (…) Financial assistance can be either bilateral or multilateral, for example, as a contribution to an international fund.
>
> (UNEP/OzL.Pro.WG.II(1)/4, p.14)

But while the general need for support was soon accepted, significant conflict emerged over what exactly this should look like. What significantly complicated the situation was the expectation that the ozone fund would be a precedent for an even larger future fund in the area of climate protection, which would mean an extensive increase in financial transfers from industrialised to developing nations (Gehring 1994, p. 293). Thus, both sides were particularly eager to push through their preferred solution, and neither was willing to give up easily.

Donors, in particular the USA and major European countries, wanted to avoid the creation of new institutions for the purpose and preferred to establish a clearing house mechanism to channel (existing) aid flows through bilateral programmes or existing institutions, in particular the World Bank (OEWG II(2)/7, Annex II). For them, a strong role of the World Bank was attractive because the weighted voting structures and its general affiliation to the West would ensure them a degree of privilege (Gehring 1994, p. 239). Further, they wanted to avoid the proliferation of funds going along with the proliferation of environmental treaties. Instead, they envisioned one common financial mechanism for all future North–South financial transfers for environmental projects with global impact (Streck 2001, p. 75).

Developing countries, on the other hand, led by China and India and supported by some smaller donor countries, aimed for the creation of a specialised fund under the authority of the protocol with a strong role for UNEP. With its close connection to the developing world, they hoped to maximise UNEP's influence as opposed to other major international organisations. Further, based on the UN one-country–one-vote principle, a fund under the authority of the protocol would put everyone on equal standing. For the same reasons, a strong role for the World Bank was attractive to the donor side, it was essentially a dealbreaker for developing countries (OEWG II(2)/7, Annex I, Benedick 1998).

It soon became clear that there was not only functional pressure to provide financial support for developing countries to enable their participation in the tightening control measures. This support further had to be structured in a way that became acceptable to the developing world. No close affiliation with the World Bank would be acceptable, and alternatives were sparse. Thus, the creation of an entirely new structure under the authority of the MOP became necessary.

Further, it became necessary to establish the fund based on a MOP decision. Even though a formal amendment would follow, this alone could not have resolved the issue. The London Amendment, which formally established the fund, would enter into force only upon deposition of the twentieth ratification but not prior to January 1992. Further, it would be binding only for parties that had deposited their instrument of ratification. Hence, there would be a transitory period of considerable length in which some members of the protocol would have ratified the Amendment while others would not have done so. During this transitory period, only those that ratified would be bound to finance the incremental costs of developing countries. If larger donors delayed their ratification, other parties would end up having to finance a larger share of the overall costs. To resolve this issue, shortly before the London MOP, an interim arrangement was discussed in the Working Group. The result was the establishment of the Interim Fund. Whereas the regular fund would be founded on the amended Montreal Protocol, i.e. on a full-fledged international treaty, the creation of the Interim Fund had to enter into force immediately for all parties at the same time and thus could only be based on a MOP decision. Thus, remarkably, this multi-million-dollar interim fund came to life based merely on a MOP decision (Gehring 1994, p. 300).

The Solution: Equal Influence for All

The (Interim) Multilateral Fund provides financial assistance to all member state parties operating under Article 5 (1) of the protocol, meaning all those that are classified under the protocol as developing countries and whose annual calculated level of consumption of controlled substances is less than 0.3 kg per capita on the date of entry into force of the protocol for it, or any time thereafter within 10 years of the date of entry into force of the protocol. This is the same group of parties that had been defined to be eligible for special support from the beginning, in particular, the grace periods on the implementation of control measures. The institutional structures of the fund represent the need to enable developing countries to join and participate in the protocol by providing them with additional financial support under the equal control of donors and recipients. This made developing countries trust that they would receive sufficient support to achieve the tightened control measures of the London Amendment.

While agreement on the need for financial assistance had been reached quickly, the question of how this assistance would be managed had been hotly disputed. On this aspect, both sides were explicit. While industrialised nations aimed to channel financial support through the World Bank, developing countries remained adamant that the fund needed to remain under the authority of the protocol itself. Eventually, aspects of both proposals were integrated into the design of a structure with minimal bureaucracy, which would incorporate the expertise of UNEP, the World Bank, UNDP, and other relevant agencies while still remaining under the policy control of the parties instead of the board of an outside organisation. Thus, the Word Bank is still integrated into the Montreal Protocol's financial mechanism by cooperating with the Executive Committee among UNEP and UNDP, and it was designated to be the administrator of the programme to finance the agreed incremental costs of developing country parties. Still, while its role is not unimportant, it remains under the authority of the Executive Committee with its equal representation of donor and recipient parties (Second Meeting of the Parties, Annex IV – Appendix IV, D., para. 15). Thus, while the World Bank has not been fully excluded, the main aim of developing countries to keep the Montreal financial mechanism away from its weighted voting structures was ultimately successful.

The donor side was clearly dissatisfied with this result. This is illustrated by their numerous attempts to relocate financial flows closer to the sphere of influence of the World Bank. In a bold attempt to present the developing country side with a fait accompli, France and West Germany, in parallel to the official Working Group negotiations, launched an initiative outside the official channels of the Montreal Protocol to create the Global Environment Facility (GEF) in cooperation with the World Bank (Benedick 1998). The GEF integrates funding for four key areas: Ozone, Climate, Water, and Biodiversity, and originally operated inside the World Bank in cooperation with UNEP and UNDP. It was created at a meeting chaired by the World Bank and attended by UNEP and UNDP, as well as 17 donor countries, in March 1990 – only months before the adoption of the London Amendment and under the exclusion of all possible recipient countries (UNEP/OzL.Pro. WG.III(2)/Inf.4, Gehring 1994, p. 306). This initiative did, however, not succeed. The Group of 77 (G77) was opposed to the GEF from the beginning and remains extremely suspicious of it until today, despite significant reforms. They opposed the dominant role of the World Bank, firstly because of the control it exercises on the financial policies of many developing countries and secondly, because many of them doubted the ability of the bank to address environmental issues and feared that the GEF would impose "green" conditionalities on World Bank loans (Streck 2001, p. 83). Although ozone protection was named as one of the key focus areas of the GEF, no formal connection to the Montreal Protocol was ever established, and the creation of the Multilateral Fund went on as planned.

Even after the London Amendment entered into force on January 1, 1992, the donor side did not relinquish right away. Although all legal instruments of the Interim Fund were drafted in a way that would allow their immediate transfer to the regular fund, a number of European donor countries once again launched various initiatives for postponement or re-negotiation (Gehring and Oberthür 1993, p. 10). At the following session of the Working Group in July 1992, the UK and the Netherlands proposed, on behalf of the EC, to integrate the Multilateral Fund into the GEF. Interestingly, the USA had now changed sides. They had secured themselves a permanent seat on the Executive Committee through informal settlement among the group of contributors, thus making the existing structures now more appealing than their originally preferred solution. The developing country side remained adamant in their opposition to the GEF and claimed that it merely provided a green façade while still assessing projects predominantly according to economic criteria (Rowlands 1993, p. 29). India and China, in particular, expressed the feeling that industrialised countries were backtracking and breaking their promises. They made it very clear that they could not accept any change in the provisions of the fund (UNEP/OzL.Pro/WG.1/7/4) and continued to vigorously struggle for its immediate establishment as planned in the London Amendment. They completely paralysed negotiations by refusing to participate in preparatory meetings for the following MOP in Copenhagen and threatened to prevent the adoption of other decisions unless agreement on the fund was reached. Only after significant disagreements a decision to establish the Multilateral Fund could finally be adopted (Gehring 1994, p. 296). Although the Working Group agreed on this by consensus, the EC did not support this and stated that, although they are not objecting, if a vote had been taken, they would have voted against (UNEP/OzL.Pro/WG.1/7/4, para. 118).

The fact that those states, that are usually the most powerful, could not achieve their preferred solution here shows clearly that functional pressure was not only present, but it was very strongly working against them. The need to convince developing countries to join the protocol was so strong that they could make significant demands based on the threat of their non-participation. If the donor side had decided to ignore their position and go ahead with their own preferred solution, it is very likely that many developing countries, India and China in particular, would have withheld their accession or refused to ratify the London Amendment with its package of strengthened control measures – with all the consequences of their non-participation outlined above. This would have severely undermined the success of the Montreal Protocol, so much so that the achievement of its aims would likely have become impossible.

Conclusion

This chapter has shown that the creation of the Multilateral Fund under the authority of the Montreal Protocol was driven by functional pressure created by a mismatch between strong and further tightening obligations and a lack of support for the achievement of these obligations for developing countries. This mismatch was due to cause significant problems for the operation of the protocol when developing countries made their participation conditional on a promise of financial assistance.

It has been shown that the aims of global equitable control of ozone-depleting substances and the eventual phase-out could only be achieved with the meaningful participation of developing countries. Developing countries, however, were adamant that they could only join if they would be sure to receive adequate financial assistance and would not risk compromising their development in order to keep up with their obligations under the protocol. The functional pressure which made their participation necessary gave them significant leverage in the negotiations over the specific institutional arrangements of the fund. And, the largest Western donor countries soon had to realise just how strong this pressure was. Numerous attempts to incorporate the fund into the sphere of influence of the World Bank, where they would be in a privileged position to decide over financial flows or sabotage its creation altogether, remained unsuccessful. Developing countries made clear that they would not accept anything less than the London compromise. Those that are usually weak could suddenly make significant demands. And, those that are usually strong found themselves in an uncomfortable position of weakness. Had they gone ahead with the creation of a fund based on their own preferences, developing countries very likely would not have accepted it, and the London Amendment would not have been ratified. As a consequence, the efforts of industrialised states under the Montreal Protocol would have been severely undermined.

In the end, the creation of the (Interim) Multilateral Fund in this highly contested form succeeded in fulfilling its purpose. Developing countries agreed to join the protocol and accept the strengthened control measures, albeit with a 10-year grace period. Members now not only include almost all industrialised countries but also the majority of developing countries, among them those whose participation was most relevant to the achievement of the protocol's aims. Most importantly, China joined the protocol in 1991, and India followed 1 year later. A dangerous loophole in international cooperation to protect the ozone layer had been effectively closed. However, some donor countries did not keep their promises in relation to the fund. By November 1992, the Interim Fund had received less than 80 per cent of 1991 contributions and less than 70 per cent of 1992 contributions. The largest part of the missing contributions was caused by the Russian Federation as a result of the economic and political turmoil of the collapse of the Soviet Union, and France was still discontent with a fund operating independently of other international institutions (Gehring 1994, pp. 305–306).

On a larger scale, the expectation that the Montreal financial arrangements would become a precedent for further environmental areas, in particular climate protection, also became a reality. Now, many MEAs include a similar hybrid financing structure with a specialised fund under their own authority and additional support opportunities through the GEF. Another impressive example, the UNFCCC and the Green Climate Fund, will be discussed in the section "The Creation of the Green Climate Fund" in Chapter 5.

Defining Developing Countries

Soon after the painstaking negotiations over the Multilateral Fund had finally found a compromise, a new and intimately related issue emerged. The Montreal Protocol, in particular in its new form after the London Amendment, clearly differentiates between the obligations and eligibility for support between industrialised and developing countries. However, it was not so clear on how the two groups themselves would be distinguished. Soon, the MOP decided to take matters into its own hands. After the adoption of an initial list of developing countries, those wanting to be listed as developing country would have to apply, and each application would be considered collectively inside the MEA.

This case is an interesting and unexpected example of authority expansion in the area of implementation decisions. The ultimate decision on which states would be considered as a developing country, with all its implications, would now be under the authority of the MOP itself. Existing lists held by other organisations were considered but explicitly not applied. With no specific criteria adopted, the institution enjoys significant leeway in making these decisions. This is relevant because the classification into one of these two groups has far-reaching consequences for the obligations of a party under the protocol. While developing country parties are eligible for grace periods and support through the Multilateral Fund, industrialised country parties have a legal obligation to provide the Multilateral Fund with its financial resources.

Why did the Montreal Protocol adopt this new capability to classify developing countries on its own instead of relying on established lists and without specifying clear and objective criteria for doing so? This chapter argues that there was functional pressure spilling over from the new integrative steps made with the London Amendment, in particular, the tightened control measures and the creation of the Multilateral Fund. Since the classification was tied closely to the obligations under the protocol, the issue necessarily had to be addressed on the collective level since a common understanding among all was needed for the protocol to function. As the relevance of the classification grew, it became clear that a flexible approach was needed that could address the issue on a case-by-case basis since existing lists or criteria were likely to upset the fragile willingness of developing countries to participate. An important catalyst for this development was the collapse of the Soviet Union, which led to the emergence of a number of newly independent states and, therefore, upset the existing order.

The Case: Defining Developing Countries

A central component of the Montreal Protocol is its concessions and support for developing countries, including, in particular, the Multilateral Fund. Specifically, the relevant provisions refer to those parties "operating under paragraph 1 of Article 5" of the protocol. This paragraph includes parties that are developing countries and "whose annual calculated level of consumption of the controlled substances is less than 0.3 kilograms per capita". Article 5 parties are allowed a 10-year grace period for compliance with the control measures of the protocol, and – under the new Article 10 of the London Amendment of 1990 – they are eligible for technology transfer and financial support through the financial mechanism and Multilateral Fund. Those not operating under this paragraph finance the Multilateral Fund through their contributions. Therefore, the classification of members into these two groups is highly important and determines their rights and obligations under the protocol. While the consumption criterion provides a clearly measurable guideline, it soon became clear that the term "developing country" was in need of clarification.

At the first MOPs in May 1989 in Helsinki, the MOP adopted decision I/12E on the clarification of the term "developing country". This decision included a list of those countries that "shall be considered developing countries for the purposes of the Protocol". This list, worked out by the Ad Hoc Working Group on Data Harmonization and its Legal Group, included the list of the countries that were members of the Group of 77 at the time, with the addition of Albania, China, Mongolia, and Namibia (130 states in total). This follows closely the blocks in which parties generally negotiated under the Montreal Protocol and the UN framework more broadly and thus was the most straightforward starting point. While the Working Group saw this list as an interim arrangement from the beginning and had recommended including a review mechanism for future amendments of this list based on economic criteria (UNEP/OzL.WG.Data.2/3/Rev.2, annex II), the final decision does not reflect this interim nature and merely adopts the list as such. Although this list relies (mostly) on a common classification, its adoption is a first step of this authority expansion. Under the assumption that it was originally seen as merely an interim arrangement, the adoption as a decision and not as a formal amendment made sense. Still, even though this list would not remain in its original form for long, it continues to build the foundation of the classification of developing countries.

At the third MOP in Nairobi in 1991, it was decided to grant the wish of Turkey to be classified as a developing country for the purposes of the Montreal Protocol, based on the fact that "Turkey is classified as a developing country by the World Bank, OECD and UNDP" (Decision III/5, para. 2). In the same decision, the MOP decides to "consider the requests by states for classification as developing countries on an individual basis as and when they come" (para. 1) and "to request the Open-ended Working Group of the parties to study and fully define the criteria which will be applied in the future in case of applications for classification as a developing country for the purpose of the Montreal Protocol" (para. 3). Based on the resulting report of the Working Group, the MOP finally decided at their fourth meeting in Copenhagen in November 1992 in Decision IV/7 to:

> (…) note that the Open-ended Working Group recommended that no criteria for future classification as a developing country for the purpose of the Montreal Protocol be adopted by the Meeting of the Parties and that the Parties should consider individually applications by Parties for classification as developing countries as and when such applications are made.

With this decision, the authority of the Montreal Protocol was significantly expanded in the area of implementation decisions. It would now decide on its own which parties would be classified as a developing country under the protocol. It was clearly decided against the adoption of objective (economic) criteria as originally proposed by the Working Group and also against the use of established lists of major international organisations like the World Bank, UNDP, and OECD, giving the protocol significant leeway in making its decision.

As a consequence, the MOP gained significant influence over which parties would be eligible for grace periods and assistance from the new Multilateral Fund and which would be obliged to pay into it. This is not only relevant for those states wishing to obtain developing country status, either because they saw themselves on the wrong list in the light of the new obligations the London Amendment would commit them to or because they had only been founded after the original list was adopted which became particularly relevant with the collapse of the Soviet Union. There were significant consequences for industrialised countries as well. With the creation of the Multilateral Fund, they faced additional financial commitments. The initial capital of the fund was US$160 million over 3 years from 1991 to 1993, to be increased up to US$240 million upon accession of additional developing countries, in particular India and China (MOP2 Annex IV – Appendix IV: Terms of reference for the Interim Multilateral Fund). If the protocol decides to move parties from industrialised to developing country status, the number of states that have to come up with this amount of money decreases, meaning a further increase in contributions for each individual industrialised member. Further, if the number of developing countries in the protocol increased, the volume of the fund would need to be adjusted to the increasing demand, meaning a further increase of contributions for industrialised parties. The recognition of this increasing demand is illustrated by the fact that when China joined the protocol as an additional developing country expecting to receive assistance, the fund was increased by US$40 million to US$ 200 million (UNEP/OzL. Pro.3/11, para. 14; Decision III/22 (4)).

Functional Pressure: The Need to Keep the Balance

The list of developing countries adopted in May 1989 was soon overtaken by history. When the Berlin Wall fell only a few months later, a number of states gained new independence inside the borders of the faltering Soviet Union. By 1996, there were 27 states on the territory that had before been covered by only 9 parties to the Montreal Protocol (Benedick 1998, p. 276). None of the newly emerging states were listed as developing countries; they simply had not existed when the list was written. Were they to join, they would have to accept the same obligations as industrialised countries. But recovering from total collapse, facing profound and tumultuous political, economic, and social changes and implementing the strict control measures of the Montreal Protocol immediately and without financial support, instead paying contributions to the fund, would be an insurmountable task (Andersen et al. 2002, pp. 145, 249; Parson 2003, p. 231).

The collapse of the Soviet Union was not only a shock to the world as a whole but it also intensified a problem which was lurking inside the structures of the newly amended Montreal Protocol. The issue of the definition of developing countries had first surfaced with Turkey's application for developing country status in 1991. With the 1990 London Amendment, which had just entered the ratification process, the classification of developing countries would now become much more relevant than it was before. Developing countries had accepted stricter control measures in exchange for the promise of financial support through the newly created Multilateral Fund, paid by those classified as industrialised nations. As such, the classification would now have financial consequences. And with a new amendment in preparation to be adopted at MOP 4 in 1994 in Copenhagen, adding even stricter control measures, financial assistance was becoming even more important. However, the issue of which parties would be considered a developing country had not been fully resolved. Technically, only those on the list adopted in 1989 – before the creation of the fund – would be eligible for support. Although the Working Group had recommended the adoption of a review mechanism for amending the list, no rules had been put into place at this point. There was no formal way for those that saw themselves on the wrong list in the light of these new obligations to change their status and no way for new states to be added to the list.

This functional interdependence between the newly created Multilateral Fund and the definition of developing countries created pressure to address the issue at the collective level. And, this pressure grew significantly with the emergence of many new states whose status was unclear. While the countries in question were relatively small, they did consume many of the controlled substances and had immense potential for growth. Turkey, the first party to change its classification, had been identified as one of the countries with the highest potential for CFC consumption growth (Kohler et al. 1987, p. 6) (see also section "The Creation of the Interim Multilateral Fund"). The estimated 1994 consumption of controlled substances of all countries with economies in transition (CEIT) was approximately equal to a quarter of global consumption in that year (Yoshida 2019, p. 273). Armenia alone consumed more CFCs than India in 1986. And, Georgia consumed more methyl bromide than China in 1991, the year this substance appeared on the radar of the parties and which has been controlled under the protocol since the Copenhagen amendment of 1994 (UNEP 2020b).

Therefore, the issue of their participation could not be neglected. If these parties would put their ratifications on hold, the success of the protocol would be in danger. A path had to be found which would enable them to join the protocol and participate according to their needs and abilities. Trying to stay afloat in the middle of a collapsing world order, the Montreal Protocol had to address this gap in its regulatory framework. This, however, turned out to be complicated. Soon, the discussion would not only be about which parties would get added to the list but also about which would be able to stay on it.

Understanding that something had to be done, the MOP decided at its third meeting in June 1991 to request the Open-ended Working Group to "study and fully define the criteria which will be applied in the future in case of applications for classification as a developing country for the purpose of the Montreal Protocol" (DecisionIII/5 para. 3). Following this decision; different options were investigated at subsequent meetings in April and July 1992. First, they concluded that no officially recognised definition of the term "developing country" existed in the UN system. The 1989 list was based on the members of the Group of 77 with the main addition of China, hence including the coalition of developing countries that stood together in negotiations about the terms of the Multilateral Fund. Major international organisations kept their own lists. The United Nations Conference on Trade and Development also used the list of members of the Group of 77. The OECD kept a list of those states that receive public development aid. UNDP lists those countries that, based on their needs, benefitted from its programmes. Finally, to determine the eligibility for funding under the GEF, the World Bank used the UNDP list and the additional criteria of a per capita income equal to or lower than 4000 US$ in 1989. Accordingly, the World Bank was the only organisation which used precise criteria (UNEP/OZL.Pro/WG.1/6/2, paras. 15–17).

Some delegations stated that "it was desirable to establish objective criteria on classification", while others "feared that any such endeavour might upset the balance achieved by the Parties to the Protocol" (UNEP/OzL.Pro/WG.1/7/4, para. 131) and advocated for a flexible approach (UNEP/OzL.Pro/WG.1/6/5 (F.). The latter group was able to prevail, and the Working Group finally concluded that it "would not make a specific recommendation on the issue" (UNEP/OzL.Pro/WG.1/7/4, para. 134). This recommendation is what Decision IV/7 of the MOP to take onto itself the authority to decide on its own case-by-case and without specific criteria was based on.

Why, if several established lists existed, was it decided not only against using one of those but against establishing any type of objective criteria despite the clear request to the Working Group to do just that? In short, why did the Montreal Protocol take the authority over the definition of developing countries into their own (subjective) hands instead of establishing objective criteria as they originally planned? The answer lies indeed in the fragile balance between industrialised and developing country parties and the need to stabilise it so the protocol could reach its aims. The existing lists were not consistent and not uncontroversial. Five countries appeared on only one of the lists, and two member states of the European Economic Community appeared on two of the lists (UNEP/WG.185/5/Rev.1, para. 18). Using one of these might have cost some states their status or add new states to the list, which not everyone would consider as deserving. Further, none of the states emerging from the Soviet Union were listed. Additionally, relying on the list of another major international organisation would relocate the control over the issue on the outside of the protocol into the hands of a different organisation which would follow its own interests, which would not necessarily match those of the Montreal Protocol. Meanwhile, Russia and other central and eastern European countries argued that their special situation had left them at least temporarily worse off than some developing countries and suggested that a third group for CEIT should be recognised to allow them temporary relief from their Multilateral Fund obligations. While they did receive some sympathy from industrialised countries, developing countries categorically rejected this proposal, and UNEP Executive Director Tolba also opposed relaxations for these countries, "fearing the opening of the floodgate that could threaten the protocol" (Benedick 1998, pp. 212, 279).

Applying objective economic criteria was also not an uncontroversial option. The only point of reference was the criterion of a GNI per capita of below US$ 4000, as used by the World Bank. Indeed, this criterion would have put all former Soviet states on the list of developing countries. Further, since the original list of developing countries was not based on economic criteria, applying this now would upset the existing status quo. On the one hand, in a preparatory meeting, one delegation stated that they were required to pay a contribution to the Interim Fund although their per capita GNI was lower than that of many developing countries, and they consumed very little of the controlled substances (UNEP/OzL.Pro.4/Prep/2, para. 24). On the other hand, a number of states on the original list had a GNI per capita above US$ 4000 since before 1989, and many more were crossing this threshold during the years the issue was discussed or expected to cross it soon. Of the original 130 developing countries on the 1989 list, 13 were already above US$ 4000 in 1989. Some of them were not operating under paragraph 1 of Article 5 and were already paying into the Multilateral Fund (Bahrain, Malta, Singapore, United Arab Emirates, and South Korea). Others, however, were in danger of losing their status if the World Bank criterion was to be applied. Among them was Saudi Arabia, which had been identified as one of the developing countries with the highest potential for CFC consumption growth. Several more states with high potential for growth were very close to the threshold and would cross it in the following years, e.g. Argentina, Brazil, (Iran, Iraq), and Mexico (Kohler et al. 1987, p. 6; World Bank 2020). Many of those had only just obtained Article 5 status (MOP 4, XIV, MOP 3, Annex X). Since states with a GNI per capita above US$ 4000 were also not eligible for GEF funding, these states would have no way of attaining funding for ozone protection projects from either the Multilateral Fund or the GEF, and they were not eligible for grace periods. This would likely have made it significantly harder for them to participate in the protocol in a meaningful way. At this point in time, Saudi Arabia had not yet joined the convention and the protocol and Argentina and Brazil had not ratified the London Amendment. If these countries were to lose their eligibility for Article 5 status, they surely would have been discouraged from accepting the strengthened measures of the London Amendment. It is likely that any other type of objective criteria which negotiators could have produced would have had similar consequences.

This problem posed a significant danger to the stability of the cooperation project as a whole. The London Amendment had, with its strengthened control measures and its extensive funding mechanism, finally established a balance between industrialised and developing country parties. It had made the terms of the cooperation project acceptable for developing countries and provided sufficient support to enable their participation without compromising their development. Many of them were finally willing to join. But now, this fragile balance was once again in danger of being upset. After they had fought together for the establishment of the Multilateral Fund, it was unlikely that they would now accept the adoption of criteria that could potentially threaten their eligibility to benefit from it. It was important to find a way to handle the situation in which no party would lose their Article 5 status that was already eligible and, at the same time, make it possible for states to be added to the list, in particular with a view to determining the status of the former Soviet states. With existing lists or clear and simple objective criteria, this would have hardly been possible. Thus, there was clear functional pressure to move the authority to define developing countries to the collective level and make these decisions on an individual case-by-case basis without the application of specified criteria.

Further, the issue of the definition of developing countries could not have been handled through a formal amendment, neither the original list nor the later decision to mandate the MOP to decide on a case-by-case basis. The reason for this is the collective nature of the issue. Since many other rules of the protocol were tied to the distinction between developing and industrialised country parties, this distinction had to be addressed collectively. Otherwise, if done as an amendment, the protocol could have run into a situation in which those that ratified such an amendment and those that did not could disagree on the classification of individual parties, leading to significant insecurity on how to handle such a situation. Thus, there was not only functional pressure to address the issue as such but also to address it collectively.

Neither existing lists nor common objective criteria could be applied without risking the achievement of the protocol's objectives. With each available option, there was the threat of upsetting key parties. The only way to resolve the situation was indeed to give the Montreal Protocol the authority to keep the original list and make individual additions and subtractions on a case-by-case basis through MOP decision. It was central to the success of the protocol that the authority over this process would be in the hands of itself and no one else, not even objective numbers.

The Solution: Expanding Flexibility

Decision IV/7 explicitly places the authority to determine, by decision, which parties are considered developing countries into the hands of the MOP. Technically, all parties to the protocol can apply for their status to be changed, and the matter will be addressed collectively. With this, the MOP is, essentially, the sole determinator of which parties pay for the Multilateral Fund and which can apply for support from it, and which can make use of grace periods on the strict reduction obligations under the protocol. Organising the classification in this way allowed the Montreal Protocol the necessary flexibility to avoid the pitfalls of existing lists and criteria analysed above.

The MOP is explicitly not bound by any type of objective criteria when making this decision, neither economic criteria nor existing lists. Although it has often referred to the lists held by the World Bank, UNDP, and OECD and the status of applicants under other international agreements when approving applications for developing country status, it has not applied these lists consistently. For example, at MOP IX in Montreal in 1997, the applications of South Africa and Moldova for developing country status were both approved, but different reasoning is provided. For South Africa, its status as a developing country at the UNDP and OECD is cited, as well as the fact that it is considered a developing country at all other MEAs to which it is a party (Decision IX/27). For Moldova, the decision is based on the World Bank classification in addition to the UNDP and OECD lists (Decision IX/26).

Further, since the decision is made on the collective level and does not require each individual state's approval, it can be made from the collective standpoint of the protocol and is not bound to each individual state's perception and relationships. This means that the MOP can decide based solely on the need to keep the balance between developing and industrialised countries that is necessary to keep the protocol up and running toward the achievement of its aims. It can take into account the delicate compromise reached with the London Amendment and its complex historical context to ensure that all parties will be classified in a way that allows them to participate in the best possible way in the protection of the ozone layer.

Conclusion

This chapter has shown that it was indeed necessary for the Montreal Protocol to decide for itself which parties it would consider a developing country, with all its implications for obligations and support. There was significant functional pressure, spilling over from the creation of the Multilateral Fund, which led to the Montreal Protocol deciding to handle the definition of developing countries on their own, flexibly and without set criteria – an additional capability which might seem small at first but had immense consequences for the obligations of some of the parties to the protocol.

An earlier expansion, namely the creation of the Multilateral Fund, created a situation which demanded even further expansion, namely the definition of developing countries at the collective level. With more countries joining and the financial aspect added to the classification as developing our industrialised country party, the question of how these two groups would be distinguished gained much relevance. All of this was taking place in a context of continuously tightening control measures, making adherence more and more costly and financial assistance all the more relevant. For many developing countries, the status they expected if they were to join the protocol and, with it, the type of obligations they would be accepting was fundamental to their willingness or even capability to participate in ozone protection. If these expectations were to be upset now, membership was likely to stagnate or even decline, and the goal of global ozone protection was in serious danger. Using existing lists held by other organisations or introducing an objective economic criterion similar to the World Bank was likely to upset the fragile balance between industrialised and developing countries, which had been tentatively achieved through the creation of the Multilateral Fund and had finally convinced many of them to join. A more flexible approach was clearly needed to ensure that the creation of the Fund would actually fulfil its purpose of buying developing country participation. Without this more flexible approach, the creation of the Fund would likely have missed its mark in this regard. Since eligibility for support from the fund was tied to the list of developing countries created at the collective level, a collective solution was necessary when this list could not provide clear guidance anymore.

In the end, this unexpected expansion of authority succeeded in stabilising the protocol on its way to achieving its aims in the middle of a collapsing world order. Decision IV/7 allowed the Montreal Protocol to integrate those of the newly emerging states that were eligible for Article 5 status without upsetting the balance established by the London Amendment. Those CEIT parties that qualified for Article 5 status under the consumption limit at this point in time (Georgia, Moldova, Kyrgyzstan, Armenia, and Turkmenistan, UNEP/OzL.Pro/WG.I/7/4, para. 132, UNEP Data) were added to the list while the others were able to manage their situation through the newly established compliance procedure, thereby not "opening the floodgate" as Executive Director Tolba had feared but still mobilising the necessary support to enable everyone to participate. All former Soviet states acceded to the Montreal Protocol and the London Amendment soon after their independence. Further, Saudi Arabia acceded in early 1993, and Argentina and Brazil ratified the London Amendment around the time Decision IV/7 was adopted. Turkey ratified the London Amendment in 1995. The Montreal Protocol was stabilised once again and could continue to operate towards the elimination of ozone-depleting substances. A threat to the achievement of its aims was once again successfully averted through the expansion of the protocol's authority.

Compliance Management

With the Montreal Protocol considered by many as the most effectively implemented MEA to date, its compliance system is highly regarded. It was the first of its kind, and, like the financial mechanism, it served as a role model for many MEAs that followed (Doelle 2021, p. 985). It is characterised by its facilitative approach, focussing on reinstating compliance over the punishment of non-compliance, which became necessary because of the specific nature of the ozone layer as a collective good.

The creation of this non-compliance procedure is an important case of authority expansion in the area of compliance management. It is puzzling since the protocol already had an extensive dispute settlement procedure at its hands through its framework convention. Although the protocol's Article 8 called for the adoption of additional mechanisms for determining non-compliance and the treatment of parties found to be in non-compliance, it was only decided later by the MEA itself that these mechanisms would consist of an entirely new procedure managed by its own subsidiary committee. It would also follow an approach that was, until then, unusual: A facilitative approach based on the idea of compliance management instead of the enforcement of the rules. As such, the Montreal Protocol did not only enhance its own authority in this area but also created a precedent for how MEAs would approach the issue of compliance in the future. Following its success, the managerial approach to non-compliance became the standard for MEAs and still persists today.

Why was the Montreal Protocol able to establish this entirely new compliance management mechanism when it already had an extensive dispute settlement procedure at its disposal? This chapter argues that the creation of this new mechanism was driven by functional pressure created by the need to ensure compliance with the strict obligations of the protocol in the light of protecting a common good. Neither relying on the existing dispute settlement procedure nor the establishment of more typical enforcement rules could have fulfilled this same purpose. The collective nature of the ozone layer made a mechanism necessary that can address compliance from a collective point of view. This also means that the mechanism needs to apply to all parties collectively, which cannot be achieved through an amendment requiring ratification but only through a MOP decision.

The Case: Introducing Compliance Management

When discussions under the Montreal Protocol began to consider how to address non-compliance, it already had a traditional dispute settlement procedure at its disposal. Article 11 of the Vienna Convention lays out extensive rules on the handling of disputes between parties concerning the interpretation or application of the convention and its protocols. Where no agreement can be reached through individual negotiation or mediation by a third party, the parties to a dispute can either accept arbitration by the MOP or submit the dispute to the International Court of Justice. If they cannot agree on one of these two options, a conciliation commission will be created, consisting of an equal number of members appointed by each party concerned and a chairman chosen jointly by the members appointed by each party. This commission renders a final and recommendatory award, which the parties are expected to consider in good faith. The final paragraph of Article 11 specifies that "the provisions of this Article shall apply with respect to any protocol except as provided in the protocol concerned", and Article 14 of the Montreal Protocol on the relationship of the protocol to the convention reinforces this paragraph by stating that "the provisions of the Convention relating to its protocols shall apply to this Protocol". As such, the Vienna dispute settlement procedure also applies to the Montreal Protocol.

Further, however, the protocol's Article 8 also requires that:

> The Parties, at their first meeting, shall consider and approve procedures and institutional mechanisms for determining non-compliance with the provisions of this Protocol and for treatment of Parties found to be in non-compliance.

The first MOP in May 1989 decided to establish an open-ended Working Group of Legal Experts to develop a proposal on the procedures and mechanisms provided for by this Article (Decision I/8). The Working Group began to discuss possible options soon after. In this early stage, deliberations were based on two proposals, one submitted by the USA and one by Finland, which both envisioned an expansion of Article 11 of the Vienna Convention. There seemed to be general consensus at this point that the new mechanism would be closely linked to the existing one (UNEP/OzL.Pro.LG.1/2). It was agreed early on that the procedure would be facilitative and non-confrontational and that its decisions should be recommendatory rather than mandatory. The Working Group also agreed that it was important to avoid the creation of an unnecessarily complex and duplicative system but, at the same time, the new procedure should not alter or weaken in any way Article 11 of the Vienna Convention (UNEP/OzL.Pro.LG.1/3, paras. 8, 9). This somewhat contradictory conclusion is a first hint at what would follow, as the new mechanism would be entirely different and not touch the old in any way.

Soon, the Working Group also began to consider the establishment of a permanent committee as the core of the new compliance procedure, the Implementation Committee (UNEP/OzL.Pro.LG.1/3, para. 10). It concluded that "there was room within the non-compliance procedure for a committee which would consider observations or reservations addressed to it through the Secretariat" (para. 17). Concerning the creation of this committee, the Secretariat received a comment from Spain which states that:

> The Compliance Committee proposed by the United States, for which there is no explicit provision either in the Convention or in the Protocol, although it invokes its Articles 2, 4 and 7, would have to be compatible with the procedures, organs and institutions provided for in Article 11 of the 1985 Vienna Convention.
>
> (UNEP/OzL.Pro.LG.1/Inf.1, emphasis added)

This quote clearly shows that the creation of the Implementation Committee as the core of a new compliance procedure went far beyond the original treaty. Article 8 was not originally understood to mean the creation of a specialised committee which would be able to exercise authority over the parties. Nevertheless, the Implementation Committee entered the draft that was submitted by the Working Group to the second MOP (UNEP/OzL.Pro. LG.1/3).

The proposal found the strongest support from the EC and many developing countries. Other parties, however, including the Nordic nations and the USA, felt that stronger measures were needed. Although their efforts were ultimately unsuccessful, they prevented the acceptance of the legal group's recommendations as the final word. Reflecting a general sense of dissatisfaction with the vagueness of the proposal, the MOPs adopted the procedure only on an interim basis and directed the Working Group to elaborate more specific terms (Benedick 1998, p. 183).

While the Working Group was busy carving out the details of the final compliance procedure, the third MOP in June 1991 decided that "Article 11 of the Vienna Convention and the Non-compliance Procedure pursuant to Article 8 of the Montreal Protocol were two distinct and separate procedures" (Decision III/2). This reflects a significant turn from early deliberations, which had focused on expanding the Article 11 dispute settlement mechanism. Indeed, the final Montreal compliance mechanism differs significantly from the Vienna dispute settlement procedure. And, as will become clear in the following parts of this chapter, this was inevitable if the protocol was to succeed.

A revised draft was adopted at MOP 4 in November 1992 (Decision IV/5.2). With this decision, an entirely new compliance procedure was established. An indicative list of possible situations of non-compliance was also discussed but never adopted. Thus, non-compliance can be understood in relation to all mandatory provisions of the protocol, including the control measures, control of trade with non-parties, and the reporting obligations under Articles 7 and 9 (Andersen et al. 2002, p. 274). Although the new Article 8 reads in some places as if referring to two contending parties, it does not operate as the traditional bilateral dispute resolution (Benedick 1998, p. 271). Instead, it addresses non-compliance from the collective perspective of the protocol. In case of non-compliance with these provisions, the procedure can be triggered in three different ways. First, a party can trigger the procedure with reservations to another party's implementation of its obligations. Second, a party can trigger the procedure with regard to itself in case it concludes that, despite having made its best efforts, it is unable to comply with its obligations. And finally, the Secretariat can trigger the procedure in case it becomes aware, when preparing its report, of possible non-compliance by any party. If non-compliance is detected, the matter will be referred to the Implementation Committee. The Implementation Committee consists of 10 out of the 198 parties to the protocol which are elected by the MOP based on equitable geographical distribution and meet twice a year. It considers and reports on any submissions it receives with a view to securing amicable solutions on the basis of respect to the provisions of the protocol. In the process, the Implementation Committee is mandated to request further information from the party in question and to undertake, upon the invitation of the party concerned, information gathering in its territory. It also maintains an exchange of information with the Executive Committee of the Multilateral Fund. The final decision on which steps will be taken to bring about full compliance with the protocol by the party in question will be made by the MOP, but the parties may request the Implementation Committee to make recommendations. They are guided by an indicative list of measures adopted as an annex to the MOP decision, which includes appropriate technical and financial assistance, issuing cautions, and, somewhat contradictory to the overall approach of the procedure, the possibility of suspension of specific rights and privileges under the protocol, which also includes technology transfer and the financial mechanism.

The establishment of this mechanism had several direct implications for the individual parties and the fulfilment of their obligations under the Montreal Protocol. It adds an entirely new system to the MEA, which puts parties' behaviour under constant collective scrutiny where there was only a mechanism to address disputes between individual parties before. Although the Montreal compliance procedure is explicitly advisory and non-judicial in nature, it is mandated to perform a number of functions which significantly enhance the authority of the protocol over its member states. Particularly interesting in this regard is the possibility of the procedure being triggered by the Secretariat if it finds that states did not follow their reporting obligations or reported insufficient numbers. With this, the detection of non-compliance now does not only fall into the hands of suspicious individual parties that might trigger the Vienna Dispute Settlement Mechanism, but it becomes an ongoing, collective process. This makes it much more likely that non-compliance will not only be detected but also handled. Further, since all possible cases of non-compliance are submitted to the Implementation Committee, and this committee will make recommendations to the parties, this small subsidiary body attains a significant amount of authority over how these cases will be handled. This is strengthened by the fact that the committee is mandated to further investigate situations of non-compliance by requesting further information and even embarking on information-gathering inside the territory of a party in question. While the final decision remains with the MOP, it relies significantly on the information and recommendations provided by the committee and the secretariat. Finally, with the Implementation Committee being asked to maintain a relationship with the Executive Committee of the Multilateral Fund of the Montreal Protocol, this means that the issues of funding and compliance now become closely intertwined. Looking at the indicative list of measures, this can mean two very different things. Likely depending on the reasons underlying a specific case, parties in non-compliance might either receive additional support to enhance their capacity to comply, or their support might be suspended as a punishment. In this context, the self-trigger option makes addressing compliance issues for states facing problems much more attractive since they can expect assistance instead of retribution. Taken together, there is now much more pressure and also more incentive for individual parties to comply with their obligations under the Montreal Protocol. As such, the creation of the Article 8 compliance procedure clearly marks an extensive case of authority expansion of the Montreal Protocol.

Functional Pressure: The Need to Address the Collective Side of Non-Compliance

Since the Vienna dispute settlement procedure does apply to the Montreal Protocol, the new Article 8 compliance mechanism might seem redundant. Indeed, British legal expert and Chairman of the Legal Working Group Patrick Szell asserted that the introduction of Article 8 into the Montreal Protocol was "merely a negotiation ploy" (p. 270) by the USA to put pressure on the EC. US chief negotiator Richard Benedick, who personally introduced the Article, however, denies this. He argues that, rather, there was general recognition in Montreal that some type of compliance procedure was absolutely essential, and the specification of its details was only postponed due to a perceived lack of time (Benedick 1998, p. 270). Whether we believe Szell or Benedick on what the original thought behind introducing Article 8 was does not take away from the fact that achieving a high level of compliance was indeed central to the success of the protocol. Scenarios developed by the Montreal Protocol's science panel showed that full worldwide compliance was essential. The 1994 scientific assessment indicated that the continuation of even relatively modest CFC production for another decade beyond the official phaseout would result in additional future cumulative ozone losses of some 9 per cent relative to the effects of full compliance (World Meteorological Organization 1994, p. 24). And, 9 per cent would make a big difference. Estimates suggested that a 1 per cent depletion of total column ozone would lead to an increase in non-melanoma skin cancer incidence of 4.8–7.6 per cent. Further, for every 1 per cent increase in ozone depletion, the incidence of melanoma would increase by up to 2 per cent, and the number of fatalities from melanoma would increase by up to 1.5 per cent. This would lead to an increase by 782,100 of melanoma cases in the USA, and the number of fatalities would increase by 187,000 for the population alive at the time and born by 2075 (EPA 1987).

The question that remains is if the Vienna Dispute Settlement Procedure would have been sufficient to create the necessary level of compliance for the institution to be effective. The answer, as will become clear in the remainder of this chapter, is most likely no. Benedick was indeed right in saying that some type of compliance procedure specific to the Montreal Protocol was essential for its success. Indeed, traditional dispute settlement, like the Article 11 mechanism, is not the most suitable approach to ensure compliance in the area of environmental protection in general. The problem is inherent in the logic of the approach to conflict. Dispute settlement is initiated by an injured state against the culpable state. This makes sense because the injured state is the one with the greatest interest in responding. However, where global commons problems such as ozone depletion are concerned, this individual logic is not useful. Here, the harms of non-compliance are widely distributed, which means that no individual state is specifically and measurably injured enough so that it would have a sufficient incentive to take action (Bodansky 2010a, p. 247). The Montreal Protocol's commitments are related to a common general objective, which is saving the common good of a healthy ozone layer. Violations in control measures or data reporting do not imply damages inflicted by one party upon another. Instead of being geographically limited, the non-performance of ozone treaty obligations affects the international community as a whole (Yoshida 2019, pp. 211–285). The collective nature of the problem means that treaty obligations, and therefore also compliance with these obligations, follow an inherently collective logic. In the case of ozone protection, this collective aspect significantly gained relevance when the parties moved from the Vienna Convention to the creation of the Montreal Protocol and it's much more specific reporting and reduction commitments. This collective, non-reciprocal nature of obligations makes it hard to identify any individually injured or specifically affected state in case of another state's non-compliance (Cardesa-Salzmann 2012, p. 113), which could trigger a traditional dispute settlement procedure. As a consequence, the provisions in MEAs regarding traditional dispute resolution have gone almost completely unused (Bodansky 2010a, p. 247). It follows that the individual dispute settlement mechanism under the Vienna Convention would not have been sufficient to stabilise cooperation and had to be supplemented (and in practice replaced) by a mechanism addressing the collective side of non-compliance (Gehring 1994, p. 319).

It became clear that a new mechanism was needed which would be able to detect and react to cases of non-compliance from the perspective of the group of member states collectively. Therefore, the authority to address non-compliance with treaty obligations had to be transferred to the collective level, to the Montreal Protocol itself as the embodiment of the common interest. There was functional pressure to not simply advance the existing dispute settlement mechanism but to create an entirely new instrument which would be able to address the collective side of compliance with ozone protection commitments. Without this step, it is likely that incidences of non-compliance could have gone unnoticed or unaddressed, creating a climate in which parties would not feel enough pressure to take their obligations seriously, and the aims of the institution could ultimately be missed. As discussed above, even a small level of non-compliance would cause significant additional harm to human health, thus going against the commitment in the Vienna Convention to take appropriate measures to avoid these adverse effects.

This is also the reason why a MOP decision was the most suitable way to establish the compliance procedure. Most delegations in the Legal Working Group indicated that they would favour the adoption of a decision "since it was important to the effectiveness of the regime that all parties would be subject to the system" (see UNEP/OzL.Pro.LG.1/3, para. 18). It is true that only a decision of the MOP, despite its shakier legal nature, would be immediately binding on all parties. The establishment through an amendment would have allowed states to delay or even decide against ratification, which would then exclude them from being subject to it. But if the new compliance procedure was to address the collective aspect of non-compliance, it needed to be binding collectively, without the possibility for individual states to opt out, making a decision the only viable option.

The Solution: Enforcement versus Management

It became clear that the traditional way of addressing non-compliance through a dispute settlement procedure would not be effective in stabilising cooperation in the area of ozone protection. An entirely new mechanism was needed which would be capable of addressing the collective side of conflicts. Once this fact was firmly established, the next question was what this mechanism would have to look like to fulfil its purpose. As described in Chapter 4, there are generally two ways of addressing non-compliance, rooted in different understandings of how non-compliance comes about: The enforcement model and the management approach.

The enforcement model is built on the assumption that states will violate treaties if the benefits of violation outweigh the costs. It follows that in order to raise compliance, treaties must raise the costs of violation by imposing sanctions. This approach seeks to influence an actor's decision-making process by increasing the costs of non-compliance so that non-compliance becomes more expensive than compliance (Bodansky 2010a, pp. 235–236). Some have mentioned, however, that it is an error to conceptualise most compliance problems as being due to intentional violations. Most states enter agreements intending to comply since compliance serves the state interests that led to the negotiation process (Jacobson and Brown Weiss 1995, p. 122). Proponents of the managerial model suggest that the sources of wrongful acts are, most often, such things as a lack of capacity or a lack of resources, or genuine differences of interpretation, rather than a case of bad faith. If so, it makes little sense to punish states for what is, in fact, a good-faith effort to comply. It follows that instead of using sticks, it might be better to lure states into compliance with carrots, including such things as technical and financial assistance (Klabbers 2008, p. 1003). The purpose of a compliance system in this context is not to punish misconduct but rather to encourage and facilitate compliance. These systems seek to promote future compliance rather than to remedy past non-compliance; they aim to influence future behaviour rather than to settle old scores. The focus is on determining the cause of non-compliance and working with the state concerned to rectify the problem. To a significant degree, the managerial approach is predominant in international environmental law. Few MEAs "punish" violators with anything more than exposure and embarrassment (Bodansky 2010a, p. 227).

Similar to traditional dispute settlement, the enforcement model would not be effective in ensuring meaningful cooperation under the Montreal Protocol or international environmental protection more generally. The enforcement model focuses on punishing those that are responsible for a breach. Damage to nature, however, is often irreparable, and no amount of punishment can restore the status quo ante. As such, identifying and punishing those that are responsible is not the most relevant task for a non-compliance procedure in environmental politics. Instead, in order to achieve effective environmental protection, the focus should rather be on the prevention of (additional) harm (Klabbers 2008, p. 1001; Koskenniemi 1992). An efficient environmental non-compliance procedure needs to recognise that it is more important to bring a non-complying party back on track than to punish that party for its failure (Andersen et al. 2002, pp. 286, 354–355). While the ozone layer can and does recover, this process is extremely slow, and every damage that is done will have consequences for human life on Earth for decades. Therefore, the prevention of further damage is much more relevant than punishment for causing the harm that has already occurred.

Further, the prospect of punishment might even discourage states from participating in an environmental institution in the first place. However, participation by as many states as possible is essential to their success in many cases. With the Montreal Protocol, there was a strong feeling that if parties felt they were being subjected to a judicial process, they would become defensive and turn in on themselves, with the result that the ozone layer would be the loser. Depriving countries in non-compliance from the "carrot" of financial support likely would have damaged their goodwill, and the Montreal Protocol might have lost them as treaty parties altogether (Andersen et al. 2002, p. 286). Since developing countries were, in parallel, fighting vigorously for a strong promise of financial support in exchange for their participation in ozone protection (see section "The Creation of the Interim Multilateral Fund"), it was unlikely that they would have accepted a cut in funding in case of non-compliance. Their main argument was, after all, that they needed support because of their lack of capacity to comply.

It became clear that a different approach was needed. The Montreal Protocol's Article 8 compliance procedure clearly follows the managerial model. Rather than deter and punish non-compliance, MEAs, with the Montreal Protocol leading the way, have taken a different track and aim to encourage and facilitate compliance. The protocol's non-compliance regime is built on the assumption that parties would act in good faith, that peer pressure would be an effective deterrent, and that when non-compliance did occur, it would most likely be the result of economic or technical problems rather than a wilful act (Bodansky 2010a, p. 230). As a consequence, an initiation of the non-compliance procedure can actually lead to increased funding for a party in non-compliance instead of cuts and sanctions. Parties do not have to fear punishment but can actually hope for assistance to get back on track soon, and further harm can be prevented. This makes participation much more agreeable for those that might find it hard to implement their obligations in all its details. Particularly interesting in this regard is the introduction of the possibility for a party to trigger the compliance procedure with regard to itself. This clause encourages member states to introduce anticipated problems into the multilateral process before they generate significant problems (Gehring 1994, pp. 317–318).

It can be concluded that there was not only functional pressure to create a new compliance mechanism as such but also that it had to be managerial in nature and avoid the pitfalls of pure enforcement in relation to international environmental protection. There was functional pressure to focus on facilitative measures which would make it possible to avoid rather than punish non-compliance and which would assist rather than discourage parties from participating and fulfilling their obligations.

The fact that non-compliance can, and regularly does, arise from reasons other than pure cost–benefit calculations and make a managerial approach to non-compliance useful and necessary was illustrated impressively by the compliance issues resulting from the collapse of the Soviet Union. The countries of Eastern and Central Europe and the former Soviet Union not classified as developing countries (for an in-depth discussion of the classification of developing countries under the Montreal Protocol, see section "Defining Developing Countries") were required to observe the control measures of other industrialised countries under Article 2 of the protocol. However, the political and economic turmoil in 1990 resulting from the end of communism and the subsequent introduction of market-based economies resulted in initial non-compliance by all these countries, with the exception of Hungary and Slovakia. It was the Russian Federation, as a successor to the Soviet Union, that proposed the option to self-trigger the compliance procedure in the first place (Andersen et al. 2002, p. 133) and made immediate use of it together with Belarus, Bulgaria, Poland, and Ukraine, to warn the MEA of their possible non-compliance even before it occurred (Decision VII/18). With the assistance of the GEF, all these countries were quickly back on track to achieving full compliance (Andersen et al. 2002, p. 282). The Montreal Protocol's managerial approach was able to help these states play their part in protecting the ozone layer. Approaching this issue through strict enforcement or even dispute settlement would have been unlikely to achieve the same results. It can be easily imagined that threatening punishment in this situation would have achieved the opposite and led the states newly emerging from the collapsed Soviet Union to abandon the cooperation altogether due to a lack of capacity to participate.

Conclusion

This chapter has shown that the creation of the Montreal compliance mechanism was driven by a functional spillover. The creation of common obligations to achieve a common goal created the demand for a common and collective compliance system. This demand could not be fulfilled by the traditional, individually focused dispute settlement mechanism of the Vienna Convention. And, it could also not be fulfilled by a mechanism focused on enforcement, which could have discouraged states from participating instead of bringing them back on track. To ensure effective cooperation to protect the ozone layer, a new, collectively focused, managerial compliance mechanism was needed.

First, with the creation of the Montreal Protocol, it became necessary to also create a new and specific system to ensure compliance with its obligations. The Vienna Dispute Settlement Mechanism alone, even in an expanded version as discussed in early meetings of the Legal Working Group, would not have been sufficient to ensure full compliance. The problem is inherent in the logic of this type of system. Traditional dispute settlement is built on the principle that an injured party will trigger the system to seek out retaliation from the offender, and a solution can be worked out through arbitration. The ozone layer is, however, a common good, and if it is damaged, the injury falls on the international community as a whole. There is not one specific party which could detect and report a breach of the rules and seek compensation. Rather, a mechanism was needed which would be able to detect and react to cases of non-compliance from the perspective of the group of member states collectively. Therefore, the authority to address non-compliance with treaty obligations had to be transferred to the collective level of the Montreal Protocol itself as the embodiment of the common interest. This also means that the adoption of the procedure as an amendment would be inappropriate since it left open the possibility that some would avoid ratification. A MOP decision which immediately becomes collectively binding was the only suitable way to adopt this collective compliance procedure.

Second, in specifying the details of this new mechanism, the focus had to be on a managerial approach over strict enforcement, on prevention rather than punishment. An enforcement-based system likely would have discouraged many states from participating in the protocol and might have caused more overall damage to the ozone layer to occur. Both traditional dispute settlement and enforcement mechanisms only take action after a breach of treaty rules has already taken place. But no amount of punishment can undo damage that is already done to the ozone layer. Further, the prospect of strict enforcement would likely discourage participation, in particular, the participation of CEIT and developing countries that felt they lacked the capacity to comply fully and immediately. Ultimately, a non-complying party is less damaging to the aims of the protocol than a non-party. As long as they are in, they can be brought back on track through the implementation assistance available under the protocol. The prospect of additional assistance in case problems occur could be a strong motivator to keep trying. Thus, it was necessary to allow struggling parties to address problems early on and assist them in getting back on track with the fulfilment of their obligations to avoid further damage before it occurs. This point also further enhances the functional relevance of the Multilateral Fund. Still, though unlikely, enforcement is also possible in situations which might require it through the indicative list of measures adopted alongside the compliance procedure.

The Montreal compliance mechanism has since impressively proven its functionality. After the collapse of the Soviet Union, it succeeded in bringing all former soviet states to compliance with the protocol's reduction obligations. Many other incidences of non-compliance have since been detected and resolved. The fact that the Montreal Protocol is regarded by many as the most effectively implemented MEA to date proves its functionality. And, its managerial approach to non-compliance has become the role model for many MEAs that followed. Meanwhile, the Vienna Dispute Settlement procedure remains dormant to this day.

References

Andersen, Stephen O.; Sarma, K. Madhava; Annan, Kofi (2002): *Protecting the Ozone Layer. The United Nations History*. London: Routledge.

Benedick, Richard Elliot (1998): *Ozone Diplomacy. New Directions in Safeguarding the Planet*. Enlarged ed. Cambridge, MA: Harvard Univ. Press.

Bodansky, Daniel (2010a): *The Art and Craft of International Environmental Law*. Cambridge, MA: Harvard University Press.

Cardesa-Salzmann, A. (2012): Constitutionalising Secondary Rules in Global Environmental Regimes. Non-Compliance Procedures and the Enforcement of Multilateral Environmental Agreements. *Journal of Environmental Law* 24 (1), pp. 103–132. DOI: 10.1093/jel/eqr022

Chapman, Sidney (1930): A Theory of Atmospheric Ozone. *Memoranda of the Royal Meteorological Society* 3, pp. 103–125.

Doelle, Meinhard (2021): Non-Compliance Procedures. In Lavanya Rajamani, Jacqueline Peel (Eds.): *The Oxford Handbook of International Environmental Law*. Second Edition. Oxford: Oxford University Press, pp. 972–987.

EPA (1987): *Protection of Stratospheric Ozone. Proposed Rule*. 52nd ed. Federal Register, 239.

EPA (2018): *Basic Ozone Layer Science*. Available online at www.epa.gov/ozone-layer-protection/basic-ozone-layer-science, checked on October 21, 2023.

Gehring, Thomas (1994): *Dynamic International Regimes. Institutions for International Environmental Governance*. Frankfurt am Main: Lang.

Gehring, Thomas; Oberthür, Sebastian (1993): The Copenhagen Meeting. *Environmental Policy and Law* 23 (1), pp. 6–12.

Jacobson, Harold K.; Brown Weiss, Edith (1995): Strengthening Compliance with International Environmental Accords. Preliminary Observations from a Collaborative Project. *Global Governance* 1 (2), pp. 119–148.

Klabbers, Jan (2008): Compliance Procedures. In Daniel Bodansky, Jutta Brunnée, Ellen Hey (Eds.): *The Oxford Handbook of International Environmental Law*. Oxford: Oxford University Press, pp. 995–1009.

Kohler, Daniel F.; Haaga, John; Camm, Frank (1987): *Projections of Consumption of Products Using Chlorofluorocarbons in Developing Countries*. Santa Monica, CA: RAND Corporation. www.rand.org/pubs/notes/N2458.html

Koskenniemi, Martti (1992): Breach of Treaty or Non-Compliance? Reflections on the Enforcement of the Montreal Protocol. *Yearbook of International Environmental Law* 3 (1), pp. 123–162. DOI: 10.1093/yiel/3.1.123

Molina, Mario J.; Rowland, F. S. (1974): Stratospheric Sink for Chlorofluoromethanes. Chlorine Atom-Catalysed Destruction of Ozone. *Nature* 249, pp. 810–812. DOI: 10.1038/249810a0

Multilateral Fund (2022): *Our Impact*. Available online at www.multilateralfund.org/our-impact, checked on October 28, 2024.

Parson, Edward (2003): *Protecting the Ozone Layer. Science and Strategy*. Oxford: Oxford University Press.

Patlis, Jason M. (1992): The Multilateral Fund of the Montreal Protocol. A Prototype for Financial Mechanisms in Protecting the Global Environment. *Cornell International Law Journal* 25 (1), pp. 181–230.

Phillips, Guy D. (1990): CFCs in the Developing Nations. A Major Economic Development Opportunity. Will the Institutions Help or Hinder? *Ambio* 19 (6/7), pp. 316–320.

Rajamani, Lavanya (2012): The Changing Fortunes of Differential Treatment in the Evolution of International Environmental Law. *International Affairs* 88 (3), pp. 605–623. DOI: 10.1111/j.1468-2346.2012.01091.x

Rosencranz, Armin; Milligan, Reina (1990): CFC Abatement. The Needs of Developing Countries. *Ambio* 19 (6/7), pp. 312–316.

Rowlands, Ian H. (1993): The Fourth Meeting of the Parties to the Montreal Protocol. Report and Reflection. *Environment: Science and Policy for Sustainable Development* 35 (6), pp. 25–34. DOI: 10.1080/00139157.1993.9929109

Rowlands, Ian H. (2008): Atmosphere and Outer Space. In Daniel Bodansky, Jutta Brunnée, Ellen Hey (Eds.): *The Oxford Handbook of International Environmental Law*. Oxford: Oxford University Press.

Streck, Charlotte (2001): The Global Environment Facility—a Role Model for International Governance? *Global Environmental Politics* 1 (2), pp. 71–94. DOI: 10.1162/152638001750336604

UNEP (2020): *All About Ozone and the Ozone Layer*. Nairobi, Kenya: UNEP. Available online at https://ozone.unep.org/ozone-and-you, checked on October 21, 2023.

UNEP (2020b): Consumption of controlled substances. Ozone Secretariat Data Centre. Available online at https://ozone.unep.org/countries/data, checked 10/21/2023

UNEP (2023): About Montreal Protocol. Available online at www.unep.org/ozonaction/who-we-are/about-montreal-protocol, checked on October 19, 2023.

World Bank (2020): World Bank Open Data. Available online at https://data.worldbank.org/, checked on October 21, 2023.

World Meteorological Organization (1994): Scientific Assessment of Ozone Depletion. WMO Global Ozone Research and Monitoring Project – Report No. 37.

Yoshida, Osamu (2019): *The International Legal Régime for the Protection of the Stratospheric Ozone Layer*. Leiden: Brill Nijhoff.

5 The UN Framework Convention on Climate Change, the Kyoto Protocol, and the Paris Agreement

Climate Change is indisputably one of the most pressing issues for human life on planet Earth. International climate regulation has expanded dramatically in the last decades, much more than in any other environmental area, starting with the adoption of the United Nations Framework Convention on Climate Change (UNFCCC) in 1992, the addition of the 1997 Kyoto Protocol and its replacement by the 2015 Paris Agreement. Annual Climate Conferences now attract more than 30,000 participants and dozens of heads of state and governments. However, in sharp contrast to the outstanding success of ozone protection under the Montreal Protocol, growing awareness of climate change has not, so far, been translated into the same level of ambitious international obligations (Rajamani and Werksman 2021, p. 492). Global efforts on climate protection have continuously been characterised by a lack of efficiency (Obergassel et al. 2020, p. 3). As of now, there is little prospect that the Paris Agreement by itself can ensure that commitments will achieve the level of reductions necessary to meet its "well below 2°C" temperature goal. And, the COVID-19 pandemic and the economic downturn it triggered has affected both the pace of the negotiations as well as the ability of states to focus on addressing the climate crisis (Rajamani and Werksman 2021, p. 509). As the most relevant Multilateral Environmental Agreement (MEA) of our time, the UNFCCC, with its Kyoto Protocol and Paris Agreement, will be the second institution to be analysed here. While another typical and long-standing MEA, its fragmented structure and limited success provide for the analysis of authority expansions in a context which differs significantly from the Montreal Protocol's comparatively smooth success. After a short introduction to the issue of climate change and the greenhouse effect and the emergence and expansion of the climate MEA, three cases of authority expansion will be investigated in detail to discover their driving forces.

Climate change is caused by the accumulation of greenhouse gases (GHGs) in the atmosphere. This intensifies what we know as the "greenhouse effect". Explained in simple terms, sunlight has a wavelength of around 0.5 microns. The Earth's atmosphere is essentially transparent to light with this wavelength so that it can pass through easily and be absorbed by the planet's

DOI: 10.4324/9781003597681-6

surface. This warms up the surface and the atmosphere, which begin to emit energy as well. They are, however, much colder than the Sun and the radiation they emit has a much higher wavelength, around 10 microns. To this type of radiation, the atmosphere is not transparent. Instead, the atmosphere absorbs it and radiates about half of it back down toward the Earth's surface, where it will be reabsorbed. This additional flow of energy from the atmosphere back down to Earth is the natural greenhouse effect. In this way, the surface is being heated not only by the radiation coming in directly from the sun but also by the radiation emitted back from the atmosphere. This is how our planet achieves its average temperature of +15°C. As long as these flows of radiation emitted and received stay constant, the temperature of the surface remains constant (Dessler and Parson 2019, pp. 13–15; Maslin 2004, p. 4).

If the atmosphere changes, however, this balance will be upset. The possibility of warming caused by changes in the makeup of the atmosphere was recognised by scientists more than a century ago, first by the Swedish chemist Svante Arrhenius in 1896 and again with more supporting evidence by the British engineer Guy Callendar in 1938. The greenhouse effect is specifically caused by complexly shaped molecules which make up only a small percentage of the atmosphere's composition. The most relevant is carbon dioxide (CO2), but also methane (CH4), and to a smaller extent, nitrous oxide (N2O), and also the chlorofluorocarbons (CFCs) and related chemicals controlled under the Montreal Protocol. If the concentrations of these GHGs in the atmosphere increase, less of the radiation emitted by the surface can escape into space. On its way up and out, it becomes trapped inside the thickening atmosphere, and more of it is eventually emitted back toward the surface (Dessler and Parson 2019, p. 17; Weart 2008, p. 4).

Over the past two centuries, human activities have sharply increased the atmospheric abundance of GHGs. There is clear proof that levels of CO2 in the atmosphere have been rising ever since the beginning of the Industrial Revolution, representing an overall 30 per cent increase. The first major source of CO2 emissions is the burning of fossil fuels – coal, oil, and natural gas – for energy production, industrial processes, and transport. The second major source is land-use change. Forests have a high capacity for storing CO2. Where they are cut down, the land is often repurposed for agriculture, urbanisation, or roads with much less storage capacities (Maslin 2004, p. 11; Dessler and Parson 2019, pp. 22–23). Methane, also known as natural gas, is emitted from rice paddies, landfills, livestock, biomass burning, and the extraction and processing of fossil fuels, as well as several natural sources. N2O is predominantly emitted from nitrogen-based fertiliser and industrial processes. And, the halocarbons are a group of synthetic chemicals used as refrigerants, solvents, and in other industrial applications (Dessler and Parson 2019, p. 23). Through all these ways, human activities slowly but steadily push the process of warming ahead every single day.

And, while the idea of a warming world might look appealing at first glimpse, it will have devastating impacts on our life on planet Earth as we know it. Surface temperature is projected to rise between 1.5 and 4.8°C by the end of the 21st century, depending on the levels of anthropogenic GHG emissions. This level of warming will lead to sea-level rise and flooding and an increasing frequency of extreme weather events, including storm surges and periods of extreme heat and droughts. Possible consequences are severe ill health and disrupted livelihoods, the breakdown of infrastructure networks and critical services, food and water insecurity, loss of rural livelihoods and reduced agricultural productivity, increasing poverty, increased displacement of people, and an increased risk of violent conflicts caused by poverty and economic shocks and loss of ecosystems, biodiversity and ecosystem goods, functions and service, altered disease vectors and trajectories, species migration (Hoffmann 2013, p. 5; IPCC 2014, pp. 10–16, 65). While the full extent of these consequences is not expected until the second half of the century, with recent models focusing in particular on the period of 2081–2100 and beyond (IPCC 2014, pp. 70–71), we are already starting to experience more and more events which are likely fuelled by early changes in the global climate.

Already in 2009, UNEP warned that the pace and scale of climate change were outstripping even the most sobering predictions. Since then, it has been continuously proven in many detail that climate change is not an issue of the far future but happening now, and that its impacts, like the melting of the ice caps, are happening even sooner than anticipated (Hoffmann 2013, p. 5).

Reducing GHG emissions to a level which effectively mitigates climate change is costly and requires far-reaching changes in the way we live. Implementing the necessary reductions poses substantial technological, economic, social, and institutional challenges, which become more costly the longer they are delayed (Höhne et al. 2020). Key measures include the reduction of energy demand and the increase of low-carbon energy supply, and the phase-out of fossil fuel power generation. This will require substantial changes in behaviour, lifestyle and culture, diet, and consumption patterns. If emission reductions do not reach the necessary level, additional investments in Carbon Dioxide Removal (CDR) technologies will become necessary. Adaptation measures can substantially reduce the risks of climate change impacts, but only to a certain limit (IPCC 2014, p. 67).

Although first suspicions that the climate was changing through human-induced emissions had been present since the late 1800s, it was not until the 1980s that the issue entered the international stage. Catalysed by a period of growing environmental awareness and the negotiations of the Montreal Protocol under intense worldwide publicity, climate change slowly started to become a matter of global interest. Initiated by recommendations of UNEP and the World Meteorological Organisation, the Intergovernmental Panel on Climate Change (IPCC) was created in 1988 to further investigate the problem from a scientific standpoint (Ivanova 2017). The IPCC published its First Assessment Report in 1990, where it concluded with certainty that:

> Emissions resulting from human activities are substantially increasing the atmospheric concentrations of the greenhouse gases (…). These increases will enhance the greenhouse effect, resulting on average in an additional warming of the Earth's surface.
>
> (IPCC 1992, p. 52)

This first report also identified many of the devastating consequences of climate change for the world's social, economic, and natural systems and concluded that a global response was needed. Clearly, the issue could not be ignored any longer.

And thus, in 1992, the UNFCCC was opened for signature. With their joining, 166 governments recognised climate change as a major global concern. Negotiated at a time when the science was not as clear or certain as it is now, and political will and saliences were still muted, the UNFCCC was not intended to be the last word on climate change but to provide a starting point for more specific commitments to be added later (Dessler and Parson 2019, pp. 27–28; Rajamani and Werksman 2021, p. 497). Article 2 states the objective of the UNFCCC and any related instruments as the:

> stabilisation of the greenhouse gas concentrations in the atmosphere at a level that would prevent dangerous anthropogenic interference with the climate system. Such a level should be achieved within a time frame sufficient to allow ecosystems to adapt naturally to climate change, to ensure that food production is not threatened and to enable economic development to proceed in a sustainable manner.

It does, however, not indicate what this level is, thus implicitly leaving this to be determined in subsequent negotiation (Rajamani and Werksman 2021, p. 498). The UNFCCC addresses not only the need to cut GHG emissions (mitigation) but also to prepare for the predicted impacts of climate change (adaptation) and to provide implementation assistance to developing countries in the form of finance, technology transfer, and capacity building. It requires all parties to develop national inventories of GHG emissions, formulate policies and measures to limit GHG emissions, and submit national communications containing information relating to their national inventories and steps taken or envisaged to implement the convention. The UNFCCC requires industrialised countries listed in Annex I to take policies and measures on GHG mitigation "with the aim of returning individually or jointly" to their 1990 levels of GHGs by "the end of the present decade", that is, by 2000. A subset of these industrialised countries listed in Annex II are also committed to providing "new and additional financial resources to meet the agreed full costs incurred by developing country parties" in complying with the reporting requirements and to "meet the agreed full incremental costs" of developing countries in implementing emissions reduction measures. Annex II parties also commit to assist particularly vulnerable developing country parties in "meeting costs of adaptation" to the adverse effects of climate change. Further, the UNFCCC establishes the institutional architecture necessary for the MEA to function and negotiations to proceed, in particular a Conference of the Parties (COP) as its supreme body, the Subsidiary Body for Scientific and Technological Advice to provide the COP with information and advice on scientific and technological matters (Article 9) and the Subsidiary Body for Implementation (SBI) to assist the COP in the assessment and review of the effective implementation of the convention (Article 10), and a financial mechanism to provide financial and technical support (Article 11).

The COP makes the decisions necessary to promote the effective implementation of the convention. Among other tasks, it periodically examines the obligations under the convention in the light of experience gained and the evolution of scientific and technological knowledge, guides the refinement of methodologies for the preparation of GHG inventories, assesses the implementation of the convention by the parties, and the overall effects of the measures taken pursuant to the convention, seeks to mobilise financial resources, establish subsidiary bodies, and is mandated to "exercise such other functions as are required for the achievement of the objective of the Convention" (Article 7.2(m)). Since the climate COP has so far failed to agree on voting rules, it adopts decisions by consensus.

The treaty also states several principles intended to guide subsequent climate policy decisions, most importantly the principle of "Common but differentiated responsibility". This principle states that all nations have an obligation to address the climate issue, but not in the same way or at the same time, and in particular that "developed-country Parties should take the lead in combating climate change and the adverse effects thereof" (Dessler and Parson 2019, p. 27). Indeed, as is the case in many international discussions, the central line of conflict would often run between the developed and the developing world.

Since its adoption, the UNFCCC has evolved in a way that is remarkable in its aspirations as well as its failures. In 1997, the UNFCCC was supplemented by the Kyoto Protocol. The Kyoto Protocol was negotiated in the context of increasing scientific certainty and a preparedness, at least among industrialised countries, to undertake specific mitigation commitments. It contains internationally negotiated legally binding GHG emissions budgets for industrialised (or Annex I) countries to apply over a 5-year "commitment period" from 2008 to 2012 and market mechanisms to reduce the costs of compliance, backed by an exceptionally strong compliance system (see section "The Kyoto Enforcement Branch"). Specifically, the protocol requires developed countries listed in its Annex B to reduce their overall emissions of a basket of GHGs "by at least 5 per cent below

1990 levels in the commitment period 2008 to 2012". Further, it identifies individual targets for each country towards this aim which range from limits of 8 per cent below 1990 levels (the European Union (EU) and its member states) to 10 per cent above 1990 levels (Iceland) (Rajamani and Werksman 2021, p. 500; Dessler and Parson 2019, p. 30). The three market mechanisms allow for additional flexibility in achieving these targets. Emissions trading (Article 17) allows parties to buy and sell emission allowances among themselves. Joint Implementation (JI, Article 6) allows industrialised country parties to earn emission reductions through projects in other industrialised countries. The Clean Development Mechanism (CDM, Article 12) allows industrialised country parties to earn emission reductions through the implementation of projects in developing countries that do not have to adhere to their own reduction targets. It is important to note, however, that it was explicitly decided that these mechanisms "shall be supplemental" and not replace domestic action (Hovi et al. 2007, p. 39) (Decision 15/CP.7).

The COP of the UNFCCC serves as the Meeting of the Parties to the Kyoto Protocol (CMP), where parties to the convention that are not parties to the protocol may participate as observers. The CMP fulfils similar functions for the protocol as the COP does for the convention. Since the rules of procedure of the COP also apply to the CMP, the CMP also decides by consensus (Article 13).

Soon, however, many became dissatisfied with the Kyoto Protocol and called for a more global approach which would include reduction targets for all countries and could convince the USA, which had not ratified the protocol, to participate. With the end of the first commitment period of the Kyoto Protocol looming, parties agreed to resume negotiations on two separate tracks, one for a second commitment period under the Kyoto Protocol, and one for "long-term cooperative action" by all nations under the Framework Convention. A second commitment period, running from 2013 to 2020 and covering only a small group of states and a fraction of global emissions, would eventually be adopted with the Doha Amendment in 2012 (Rajamani and Werksman 2021, p. 496). By then, however, the end of the Kyoto Protocol had already been cemented. At the following meetings, negotiators developed a new approach to international climate governance, which culminated in the adoption of the Paris Agreement in 2015 (Dessler and Parson 2019, p. 31).

The Paris Agreement was hailed as a historic achievement, the "world's greatest diplomatic success" (Harvey 2015). It was adopted in 2015, entered into force rapidly in 2016, and came into full operation when its parties' nationally determined contributions (NDCs) took effect in 2020. It was ratified by 193 nations, including the USA and many developing countries. While the USA shortly left the Paris Agreement under President Trump, his successor Biden quickly re-entered. The COP of the UNFCCC also serves as the meeting of the Parties to the Paris Agreement (CMA) in a similar way as it does for the Kyoto Protocol (Article 16). The Paris Agreement essentially replaces the Kyoto Protocol as the source of specific commitments to the general framework of the convention (Rajamani and Werksman 2021, p. 502). Article 2.1 identifies the purpose of the Agreement as:

> [h]olding the increase in the global average temperature to well below 2°C above pre-industrial levels and pursuing efforts to limit the temperature increase to 1.5°C above pre-industrial levels.

Further, the agreement aims to increase adaptation and climate resilience and to make finance flows "consistent with a pathway towards low greenhouse gas emissions and climate-resilient development". In order to achieve the long-term temperature goal, Article 4.1 requires parties to "aim" for global peaking of GHG emissions as soon as possible and to undertake "rapid reductions" thereafter to achieve a balance between GHG emissions by sources and removals by sinks ("climate" or "carbon" neutrality) in the second half of this century (Rajamani and Werksman 2021, p. 502).

The crucial difference between Kyoto and Paris lies in the nature of targets. In contrast to the internationally negotiated and legally binding targets of the Kyoto Protocol, under the Paris Agreement, all member states, industrialised and developing countries alike, set their own commitments in their NDCs. Parties to the Paris Agreement are obliged to submit NDCs but not to achieve them, Article 4 (2) only requires that they "intend" to achieve them. Parties are, however, expected to ensure that every successive NDC represents a "progression" on its last, its "highest possible ambition" and its "common but differentiated responsibilities and respective capabilities" "in light of different national circumstances" (Rajamani and Werksman 2021, pp. 503–504). Although the Paris Agreement offers parties considerable flexibility in determining the content of their NDCs, it seeks to restrain and direct the discretion they have and generate the necessary pressure to ratchet up their NDCs by, inter alia, requiring parties to provide information to enhance the clarity, transparency, and understanding of their NDCs, as well as track progress in implementing and achieving their NDCs through an extensive transparency mechanism.

The success of this approach has, so far, been limited. Current NDCs will limit warming to about 2.4°C. Counting only policies that are presently in place, the projected result is 2.7°C (Climate Action Tracker 2023). The UNEP Emissions Gap Report (2019) finds that "[c]ountries must increase their NDC ambitions threefold to achieve the well below 2°C goal and more than fivefold to achieve the 1.5°C goal". This gap in mitigation ambition is matched by gaps in adaptation and support. A need for as much as US$ 1.5 trillion annually is estimated, only a fraction of which is now in place. And gaps in mitigation will have a direct effect on the extent of the need for adaptation, which has largely been left to national governments, many under-resourced (Rajamani and Werksman 2021, pp. 509–510).

The scale of the challenge of combatting climate change – requiring "rapid and far-reaching transitions" across the global economy "unprecedented in terms of scale" – is immense, and achieving the necessary levels of emissions reductions may require the use of "CDR" or negative emissions technologies, solar radiation management and ocean fertilisation. Indeed, the IPCC's 1.5°C report includes these technologies in many of its scenarios already. The regulation of these technologies has, so far, been neglected under international law and remains untouched by the Paris Agreement. Further, dramatic shifts in political will are required, in particular among major emitters such as the USA and Brazil. Whether this will be possible in time to make use of the narrow window of opportunity for effective climate action remains to be seen (Rajamani and Werksman 2021, p. 510).

Compared with the Montreal Protocol, the climate MEA went through significantly less authority expansions. Only a small number of – nevertheless very interesting and maybe even more striking – instances can be detected under the convention and the Kyoto Protocol, under Paris, no major cases can be observed so far. Two reasons seem evident: First, when drafting the UNFCCC and the Kyoto Protocol in 1992 and 1997, states had already learned from their experiences with the Montreal Protocol and thus were able to anticipate functional needs, e.g. the creation of the Montreal Multilateral Fund (MLF) in 1991 was likely taken into account when drafting Article 11 of the UNFCCC which included an extensive financial mechanism already in its first draft. Second, the Paris Agreement is a very new arrangement but heavily relies on the structures of the UNFCCC and the Kyoto Protocol, which have been in existence for more than 20 years. Most functional gaps have long been closed. At the same time, the Paris Agreement is a new beginning, allowing states to remodel the institutions to their wishes. It has only recently started to fully operate, meaning that not all gaps and opportunities might have become apparent just yet. Nevertheless, three extensive cases of authority expansion with major relevance for member state obligations can be identified and will be analysed in detail in the following chapters to explain why they happened.

First, the Kyoto Protocol established an impressively strong compliance system, including a quasi-judicial enforcement branch. The creation of this committee significantly expanded the authority of the MEA to punish non-compliance through a small body of experts with limited political oversight. The section "The Kyoto Enforcement Branch" argues that the creation of this mechanism was necessary to close gaps in the at this point in time firmly established managerial approach, which were insufficient to cover the specific nature of the Kyoto commitments. Since the Kyoto Protocol included specific emission reduction targets only for Annex I (industrialised country) parties that are themselves the donors of financial aid and the drivers of technological innovation, a purely managerial approach focused on facilitating access to these forms of assistance would not be effective. Further, since the Kyoto targets were hard and costly to achieve, a strong compliance mechanism was needed to give parties the necessary level of confidence that others would adhere to them so they would do so themselves.

Second, the creation of the Adaptation Fund together with its Adaptation Fund Board (AFB), governed by a majority of developing country parties, was the first step to move the MEAs financial mechanism away from the Global Environment Facility (GEF) as its only operating entity and under the authority of the CMP. The section "The Adaptation Fund Board" argues that the creation of this peculiar structure was not driven by functional pressure but by the exceptional force with which developing countries fought for more control over adaptation finance – an issue that, for them, was a matter of survival while for industrialised countries, it was merely another part of development aid. Two characteristics of the Adaptation Fund worked in their favour. The fund does not primarily depend on industrialised country party donations, but instead, its main resource is a share of proceeds from the CDM. Further, the fund was explicitly established under the Kyoto Protocol, which does not entail the same membership as the UNFCCC and thus the existing financial mechanism, which strengthened the argument for a separate structure.

Third, the creation of the Green Climate Fund (GCF) as the second operating entity of the financial mechanism represents an exceptional shift of authority by moving the financial mechanism even further away from the donor-dominated GEF which had previously been its only operating entity. The section "The Creation of the Green Climate Fund" argues that the creation of the GCF was necessary to make the adoption of the Paris Agreement possible. The fact that the Kyoto Protocol only included specific emission targets for industrialised country parties had created a gap which threatened the success of climate protection efforts. Global emissions did not actually decline but merely move to those parts of the world where no reduction targets were in place (this is often referred to as "carbon leakage"). To convince developing country parties to accept costly targets of their own and make effective climate protection possible, they needed something in return. The creation of the GCF essentially became part of a package deal which gave developing country parties new financial assistance they could equally control in exchange for their acceptance of emission targets.

The Kyoto Enforcement Branch

The Kyoto Protocol's compliance system constitutes a landmark in global environmental governance. It consists of an independent body, which is unique for MEAs in its objective to enforce as well as facilitate compliance (Lefeber and Oberthür 2012, p. 77). The Kyoto compliance system has been the only one to date to experiment with this combination through two separate branches of an integrated committee (Doelle 2021). Its most distinguishing features in this regard are the toughness of its enforcement consequences and the ability of a small body of experts to impose these consequences with only limited political oversight (Werksman 2006, p. 19)

The creation of the Kyoto compliance committee is a particularly striking authority expansion in the area of enforcement. Its quasi-judicial enforcement branch disciplines industrialised country parties through sanctions that are exceptionally hard for the realm of environmental politics. This includes not only an obligation to prepare additional plans and reports but also the exclusion from the Kyoto flexibility mechanisms and even the application of penalties on emission targets in subsequent commitment periods. While the legal status of these rules is built on thin ice, they are formulated and have been applied in a binding way.

This chapter investigates why this exceptionally strong compliance mechanism emerged. It argues that the key lies in the specific nature of the Kyoto commitments, which created functional pressure for the inclusion of the enforcement approach in addition to the facilitative approach that had become standard for MEAs with the Montreal Protocol. Two central aspects stand out. First is the fact that the Kyoto Protocol includes specific emission reduction targets only for Annex I (industrialised country) parties. Since they are the donors of financial aid and drivers of technological innovation, a purely managerial approach focused on facilitating access to these forms of assistance would not have been effective. Second, the Kyoto targets were hard and costly to achieve. A strong compliance mechanism was needed to provide confidence for parties that they would not fall victim to free riders – or they would not make this substantial effort themselves. A purely managerial compliance system could not have resolved this issue. Further, adopting the needed enforcement system in the form of an amendment would have created a significant loophole, giving states the opportunity to avoid subjecting themselves to its exceptionally strict rules by postponing or avoiding to ratify. This would have defied the logic of providing confidence that everyone would make the effort. Thus, an informal expansion through a COP/CMP decision was the only feasible way to resolve this issue.

The Case: Introducing Enforcement

The compliance committee was one of the last pieces of the original Kyoto machinery to come to life. It represents over a decade of tedious negotiations to tailor a mechanism that could fit the specific nature of the Kyoto targets and would align with the requirements of its Article 18:

> The Conference of the Parties serving as the meeting of the Parties to this Protocol shall, at its first session, approve appropriate and effective procedures and mechanisms to determine and to address cases of non-compliance with the provisions of this Protocol, including through the development of an indicative list of consequences, taking into account the cause, type, degree and frequency of non-compliance. Any procedures and mechanisms under this Article entailing binding consequences shall be adopted by means of an amendment to this Protocol.

The negotiations of the Kyoto compliance system started at COP 4 with the adoption of the Buenos Aires Programme of Action, which sets an agenda for the preparation for entry into force of the Kyoto Protocol. It established a Joint Working Group (JWG) on Compliance which was tasked to "develop procedures by which compliance with obligations under the Kyoto Protocol should be addressed" (FCCC/CP/1998/16/Add.1, Annex II). The JWG's mandate did not mention Article 18 but instead referred to all compliance-related elements of the protocol. The motivation behind this was likely to ensure that the JWG took into account compliance issues outside Article 18, such as rules related to the Kyoto mechanisms. Further, it represents an effort by some delegations to move around the requirement stated in the last sentence of the Article, which links binding consequences to the adoption of a formal amendment (Werksman 2006, p. 33).

The issue of compliance would soon prove to be a contentious one. The Montreal Protocol had set a precedent for MEAs with its managerial approach. As the Kyoto protocol evolved, however, attitudes shifted back towards tougher enforcement procedures. The central negotiating dynamic developed inside the group of Annex I states, those that had accepted specific and binding emission reduction targets and to whom tougher consequences would apply in case they missed the mark. The USA (before it was decided that they would not ratify the protocol), Canada, New Zealand, Switzerland, and the EU shared a common commitment to a strong and effective compliance system characterised by an enforcement function. Japan, Australia, and the Russian Federation, on the other hand, rejected the need for a heavy hand. Until the final stages of the negotiations, they remained unconvinced that the Kyoto Protocol's targets and mechanisms were sufficiently different to justify a revolutionary new approach and remained committed to the idea of good faith and non-binding consequences. Developing countries negotiated as a bloc through their traditional caucus of the G77 and China. Their main concern was that the compliance system would reflect the same "common but differentiated" design as the protocol's targets and they insisted on two design elements specifically: a strong enforcement system and one applicable exclusively to Annex I parties (Werksman 2006, pp. 24–26).

At COP 6, part 2, in Bonn, negotiators reached consensus on the need for tough consequences and agreed on terms that described the application of these consequences in clear and mandatory terms. Delegations reached COP 7 in Marrakesh with the main task of translating these political agreements into procedures, mechanisms, and institutions. One issue, however, would not be resolved in Marrakesh. According to Article 18 of the Kyoto Protocol, any procedures and mechanisms entailing legally binding consequences must be adopted by means of an amendment. So far, negotiators had failed to agree on how to bring the compliance system into force. The Marrakesh Accords did not resolve this fundamental dispute (Werksman 2006, pp. 24–32). As with many other aspects of the Marrakesh Accords, the issue was deferred, in neutral terms, to the CMP by "[n]oting that it is the prerogative of the Conference of the Parties serving as the Meeting of the Parties to the Kyoto Protocol to decide on the legal form of the procedures and mechanisms relating to compliance" (Marrakesh Accords, Decision 24/CP.7, FCCC/CP/2001/13/Add.3). Indeed, in its Decision 27/CMP.1, the CMP decided to:

> (...) commence consideration of the issue of an amendment to the Kyoto Protocol in respect of procedures and mechanisms relating to compliance in terms of Article 18, with a view to making a decision by the third session of the Conference of the Parties serving as the meeting of the Parties to the Kyoto Protocol.

The CMP also requested the SBI to consider the issue at its 24th session. And, from this point on, the discussion of an amendment in the sense of Article 18 slowly began to fizzle out. The SBI continuously postponed the issue throughout the following years until a conclusion was finally made at SBI 37 in November 2012 (FCCC/SBI/2012/33):

> 140. The SBI concluded that no further discussion was required under this agenda item and deemed its consideration of this agenda item completed.
>
> 141. On the basis of the above, the SBI recommended that the CMP conclude its consideration of the proposal.

By this time, the first commitment period of the Kyoto Protocol had already ended, the Doha Amendment on a second commitment period included only a handful of parties, and negotiations on a new long-term perspective for climate protection, which would eventually result in the adoption of the Paris Agreement, had long since begun – making the question of an amendment in the sense of Article 18 overtaken by history. Thus, the features of the Kyoto compliance mechanism remain with the Marrakesh Accords (Decision 24/CP.7), an extensive document laying out many specifications of the protocol, but which is adopted only by means of COP decision.

The Kyoto compliance system integrates soft and hard approaches to non-compliance into one committee consisting of a facilitative branch and an enforcement branch. It combines two fundamentally different ways of approaching non-compliance – management and enforcement (see Chapter 2). The facilitative branch is "responsible for providing advice and facilitation to parties in implementing the Protocol, and for promoting compliance by Parties with their commitments under the Protocol" (Decision 24/CP.7, Annex, IV (4)). This specifically includes an early warning function with respect to questions of implementation regarding: (a) emission targets prior to the end of the relevant commitment period and (b) methodological and reporting requirements prior to the first commitment period. With respect to any question of implementation addressed by it, the facilitative branch is to apply a mix of consequences that could be described as "soft" (Lefeber and Oberthür 2012, p. 81). It can formulate recommendations and facilitate financial and technical assistance, including technology transfer and capacity building from sources other than those established under the convention and the protocol for developing countries. Finally, it is specifically responsible for reviewing the parties' reports on their use of the flexibility mechanisms and how it supplements (and not replaces) domestic action (Hovi et al. 2007, p. 439).

While the facilitative branch can address all parties in relation to all commitments of the protocol, the enforcement branch has exclusive jurisdiction over the specified, legally binding and target-related commitments of the Annex I (industrialised) countries (Werksman 2006, p. 19). The enforcement branch is responsible for addressing potential cases of non-compliance by industrialised countries with: (a) their emission-limitation or reduction commitments under Article 3.1 of the protocol (their emission targets); (b) the key methodological and reporting requirements under Articles 5.1 and 5.2 and 7.1 and 7.4; and (c) the eligibility requirements for participation in the carbon-market mechanisms under Articles 6 (Joint Implementation), 12 (CDM), and 17 (Emission Trading) (Lefeber and Oberthür 2012, p. 80). The technical compliance of the Annex I parties is reviewed annually by Expert Review Teams (ERTs), which can raise questions of implementation in relation to the parties' reports. Further, any party can raise a question of implementation with regard to itself or with regard to any other party (Werksman 2006, p. 19). The practice of the compliance system confirms the broader experience with comparable mechanisms under other MEAs and international institutions in that the Kyoto's non-state ERT trigger has proven crucial (Lefeber and Oberthür 2012, p. 86). In case non-compliance is detected, the enforcement branch is authorised to apply punitive (or "hard") "consequences" to countries that fail to comply with their Kyoto obligations (Hovi et al. 2007, p. 439). These consequences include findings of non-compliance, the preparation of compliance plans, and suspension of privileges under the Kyoto Protocol. In case of failure to reach emission reduction targets, the application of a penalty in the next commitment period is automatic (Doelle 2021). The enforcement branch has little discretion in the application of the consequences at its disposal (Lefeber and Oberthür 2012, p. 95).

Each branch consists of ten members and ten alternate members. The chairpersons and the vice-chairpersons of the branches together form a four-member bureau which allocates questions of implementation to the appropriate branch. Both branches are composed according to the same formula. That is one member from each of the five UN regional groups, one nominated by a small island developing (SID) country, two nominated by developed countries (that is, parties listed in Annex I of the UNFCCC), and two nominated by developing countries (non-Annex I parties). Thus, in effect, 60 per cent of the members of the committee and of each of its branches are nominated by developing countries (Lefeber and Oberthür 2012, p. 80). The equitable geographical representation in both branches was essentially a quid pro quo to avoid financial penalties as an enforcement consequence for non-compliance of Annex I parties with their commitments (Werksman 2006, p. 28). Driven by the desire to minimise political interference, independence, and impartiality are emphasised. Members and alternates "shall serve in their individual capacities". The rules of procedure further specify that both members and alternates shall "act in an independent and impartial manner and avoid real or apparent conflicts of interest". Both the facilitative and the enforcement branch are required to aim for consensus when making decisions but may, as a last resort, adopt a decision by a three-quarters majority. Decisions of the enforcement branch require, in addition, a simple majority among the members nominated by developed countries and a simple majority among the members nominated by developing countries. This increases the risk of stalemate since the opposition of two members nominated by developed countries would suffice to block a decision. Further accentuating the objective to shield the quasi-judicial decision-making of the committee from political interference, the CMP is not required to confirm the decisions of the branches on questions of implementation and, in contrast with the compliance systems of several other MEAs, cannot overrule such decisions. The CMP has delegated final decision-making authority on questions of implementation to the branches and is itself limited to considering the committee's reports, adopting further rules of procedure, providing general policy guidance, adopting decisions on proposals on administrative and budgetary matters, and deciding appeals (Lefeber and Oberthür 2012, p. 85).

While this structure is uniquely strong in comparison to most other MEAs, it also suffers from an obvious weakness. Technically, it is not legally binding. As stated clearly in Article 18 of the Kyoto Protocol, it can be made so only through an amendment, which requires a three-fourth majority vote by the CMP. Even if such an amendment is passed, the compliance system becomes binding only on those countries that ratify the amendment (Hovi et al. 2007, p. 446). Thus, the consequences outlined in the procedures and mechanisms of the compliance procedure are, strictly speaking, not mandatory, although they are formulated in mandatory terms ("shall"). The question can be asked, however, if their adoption in legally binding form would really make a significant difference to their impact. In the practice of the UNFCCC, other decisions have been adopted that are not formally binding but that contain an array of mandatory terms. Notably, many of the decisions compiled in the Marrakesh Accords, including the decisions governing the Kyoto mechanisms, contain such terms. Many MEAs have used such "mandatory" but technically non-binding devices very successfully. If a party truly wishes to resist the imposition of non-compliance consequences, it would likely do so whether they are formally binding or not (Brunnée 2003, p. 278). Indeed, the enforcement branch has proven its ability to effectively address and resolve cases of non-compliance within the framework of applicable rules. As of now, the enforcement branch has addressed questions of implementation with respect to Greece, Croatia, Bulgaria, Romania, Ukraine, Lithuania, Slovakia, and

Monaco, all of which made successful efforts to come back into compliance. A procedure in relation to Kazakhstan is ongoing since 2019. Canada was able to resolve a question of implementation before a preliminary decision was adopted. The compliance procedure has also constituted an important incentive for parties to avoid compliance problems and to try to resolve problems during the ERT process. No question of implementation has arisen from ERTs with respect to the reporting deadlines regarding the initial report and subsequent annual inventory submissions. This pattern contrasts with the more common disregard of reporting deadlines for national communications under the UNFCCC and the protocol (Lefeber and Oberthür 2012, p. 98; UNFCCC 2023).

A second major weakness, however, can be seen in the lack of mobilisation of its facilitation function. While the enforcement branch has effectively addressed a number of cases, the ability of the facilitative branch to address issues of potential non-compliance before they come under the mandate of the enforcement branch (e.g. through being able to act on early warning signals) has been hampered by the lack of effective triggers to support this function (Dagnet and Northrop 2017, p. 340). The facilitative branch has not addressed any questions of implementation in substantive proceedings. It may be that part of the facilitative function of the overall system is being effectively discharged through the ERT process. However, the facilitative branch has not been able to address the looming non-compliance of Canada with its emission target. In December 2011, Canada publicly declared that it did not plan to meet its emission target and left the agreement. This incident appears to fall squarely under the "early warning" function, which, however, failed to react in time. Only parties can trigger the early warning function (through the self-trigger or the party-to-party trigger), leaving no basis for the ERTs to indicate in their reports a question of implementation that relates to potential or likely future non-compliance. No party has so far raised a question of implementation of this kind. The resulting inability of the facilitative branch to address, let alone resolve, Canada's non-compliance has led to heavy criticism of the compliance system as a whole, including in the corridors during conferences held pursuant to the UNFCCC and the protocol (Lefeber and Oberthür 2012, p. 99).

A third major problem is the fact that the effectiveness of the compliance mechanism and the consequences it can apply depend on the adoption of subsequent commitment periods. Reducing emissions further (with a penalty rate for the deficit) will be mandatory only for countries that ratify an amendment containing such a second commitment period (Hovi et al. 2007, p. 446). The fact that the second commitment period could not keep up with the first and the Kyoto Protocol soon met its demise and was replaced with the Paris Agreement meant that the Kyoto compliance system never got the chance to realise its full potential.

Still, for the time the Kyoto Protocol was up and running, this exceptionally strong and comprehensive compliance mechanism had significant consequences for parties that go beyond the typical reduction of authority that comes along with the creation of a compliance system. Most MEAs to date include a compliance mechanism which exclusively follows a managerial approach similar to the compliance committee of the Montreal Protocol (see section "Compliance Management" in Chapter 4). The addition of an enforcement branch under the Kyoto Protocol significantly tightens the responsibilities of Annex I countries in addition to the already strict and binding emission targets. While developing countries did not fall under the jurisdiction of the enforcement branch, they were aware, at the time, of the possibility that they would be eventually if subsequent commitment periods would be successfully implemented. The "hard" consequences the enforcement branch is not just mandated but obligated ("shall") to apply in instances of non-compliance are an exceptionally strong expansion of the MEA's authority over its Annex I parties. The quasi-judicial nature of the compliance committee, with its members serving in an individual capacity, the ERT trigger, its limited opportunities for appeal and very restricted opportunities for the CMP to interfere make the Kyoto compliance committee exceptionally independent and isolate it from party influence. From the perspective of Annex I parties, the composition of the committee and the branches with essentially a developing country majority adds to this significantly. It is not only those that are under the branch's jurisdiction among themselves that make the decisions but they come under the scrutiny of all parties to the protocol. This is an exceptional and, so far, unprecedented expansion of MEA authority. In this context, the fact that the consequences applied by the enforcement branch are essentially not legally binding seems to be more of a technicality than a significant issue. It is even more striking that this unexpectedly strong mechanism could emerge in the form of a COP decision despite the explicit requirement of an amendment in the enabling Article of the protocol.

However, the case of Canada might curb the enthusiasm about the effectiveness of this mechanism. Instead of facing the consequences of its failure, Canada decided to leave the agreement altogether. The exceptional strictness of the compliance mechanism might also have contributed to the demise of the Kyoto Protocol, even though this issue should not be seen in isolation from the broader issues at play in this context. Still, the fact that the Kyoto compliance mechanism remains the only one of its kind does not diminish the excitement of the investigation. Quite the opposite is true. Why did such a unique mechanism come about in this specific constellation? The remainder of this chapter will examine why this exceptionally strong compliance system came about.

Functional Pressure: The Need to Discipline the Industrialised World

The starting point for an explanation of this exceptionally strong compliance mechanism lies within discerning the central differences to other MEAs, in particular the Montreal Protocol, as the role model for a softer, facilitative approach. These differences will lead the way to uncover why a similar approach alone would not have been sufficient to ensure compliance with the special structure of the Kyoto targets. The commitments of the Kyoto Protocol are special in two different ways. First, under the Kyoto Protocol, only Annex I parties have to adhere to specific legally binding emission reduction targets. This differs significantly from the Montreal Protocol, where all parties had to phase down or out the relevant substances, although grace periods for developing countries applied. Second, compliance under Kyoto was expected to be much more economically and politically expensive than under Montreal (Werksman 2006, p. 23). Phasing out ozone-depleting substances was a much easier task than lowering emissions of CO2, which dominate our everyday lives much more deeply. The Montreal Protocol was expected to produce a net benefit, but the Kyoto Protocol would produce a significant loss. While it is hard to calculate and compare the cost for both agreements in financial terms, and any attempt to do so will always be controversial in one way or another, a general idea can be presented. Based on Sunstein (2007) and Nordhaus and Boyer (2000), the numbers in Table 5.1 can be used for a rough impression of the numbers negotiators had in mind at the time:

These two central differences are significant for the design of the Kyoto compliance mechanism. Under the Montreal Protocol, non-compliance was expected to be predominantly a problem of capacity, of developing countries lacking the necessary means to achieve their targets. Under Kyoto, this is very different. Here, non-compliance is an issue of the industrialised world, of those that are the donors of financial aid and drivers of technological innovation. As discussed in more detail in Chapter 2, non-compliance can be addressed through two different approaches – management or enforcement, depending on what is assumed to be the underlying problem causing non-compliance. A managerial compliance mechanism aims to lower the cost of compliance for those that are struggling to mobilise the resources necessary to achieve their commitments on their own. The Montreal system has worked well in this regard since it is focused on solving the compliance problems of those countries that are eligible to receive financial and technical assistance. This approach, however, is only appropriate where non-compliance is indeed attributable to a lack of resources. Under Kyoto, it is the donors of financial

Table 5.1 Global cost/benefit of the Montreal and Kyoto Protocol

	Montreal Protocol	*Kyoto Protocol*
Monetised Cost	235 billion US$	338 billion US$
Monetised Benefit	1.131 billion US$	96 billion US$
Net Cost/Benefit	+896 billion US$	-242 billion US$

Source: Based on Nordhaus and Boyer (2000) and Sunstein (2007).

means and drivers of technological innovation themselves that will be faced with compliance issues. Thus, traditional managerial compliance incentives such as transfers of financial and technological resources make little sense as a response to the parties most likely to exceed their Kyoto caps (Werksman 2006, pp. 22–23). A purely managerial approach to non-compliance would clearly be ineffective.

Instead of a lack of capacity, non-compliance under the Kyoto Protocol will likely result from attempts to free-ride because of the high costs of achieving the Kyoto targets and distrust that one might fall victim to the free-riding of others. Since targets are hard to achieve, states need to be able to trust that everyone is actually making the effort to be willing to make the effort themselves. In the light of competitiveness concerns under this much less enticing cost–benefit structure of Kyoto, delegations from North and South were calling for a "strong and effective" compliance system that would "provide parties with certainty and confidence" (Werksman 2006, p. 23). This is the type of non-compliance that resonates more closely with the enforcement model, which is built on the assumption that states will violate treaties if the benefits of violation outweigh the cost. It follows that in order to raise compliance, treaties must raise the costs of violation so that non-compliance becomes more expensive than compliance. This can only be provided by an enforcement-based system.

A further aspect that is closely connected is the functionality of the Kyoto market mechanisms. Carbon offsets and allowances only take on a marketable value if they are in demand, and they will only be in demand if parties take their commitments seriously. A purely facilitative approach to non-compliance would not answer parties' concerns about the need to ensure that all parties would pull their weight and provide sufficient confidence to investors (Werksman 1998). This close design link between the strength of the compliance procedure and the effective operation of the protocol's market-based mechanisms was key to building a consensus around a strong compliance system. Even those Annex I delegations that remained ambivalent toward tough enforcement consequences found it difficult to balance a soft approach with their enthusiasm for the protocol's market mechanisms (Werksman 2006, p. 23). As will be discussed in depth in section "The Adaptation Fund Board", at least the CDM was likely not completely necessary for the operation of the Kyoto Protocol. It is impossible to say if this is true for all three flexibility mechanisms combined. It seems likely that the achievement of the aims of the MEA still could have been technically possible; however, it required an effort of a magnitude which was hardly acceptable for the parties involved. After all, the collective emission reduction target for Annex I countries had originally been set lower and was raised from 5 to 6 per cent only in exchange for the adoption of criteria in other areas, including the flexibility mechanisms (Bettelli et al. 1997, p. 8). Thus, a lack of confidence in the functionality of carbon markets could have become a serious problem for the functionality of the protocol as a whole.

It became clear that a purely managerial approach would not have been appropriate to ensure compliance under the Kyoto Protocol and, thinking further, to ensure that the Kyoto Protocol as a whole could run in a meaningful way. This includes the daily operations under the protocol and the operation of the flexibility mechanisms, as well as the achievement of the parties' emission reduction targets, which represent the core of the agreement. There was functional pressure to include enforcement elements into the Kyoto compliance system. Those that were going to face compliance issues did not need financial or technical support since they themselves were the donors and innovators. Instead, a system was needed, which would make non-compliance costly. Compliance would be expensive, so non-compliance needed to be even more expensive. Only then could parties trust that others were likely to adhere to their targets, and so they would adhere to theirs. Further, a strict compliance procedure was necessary to ensure that carbon offsets and allowances would take on a marketable value. While the functional necessity of this last point cannot be irrefutably determined, it unquestionably adds to the points made before.

Fundamentally, the enforcement system can only provide these required functions if it becomes immediately binding on all members without exception. Even though Article 18 requires an amendment for the adoption of binding consequences, this clashes with the need to provide the confidence described above. The adoption of an amendment requires a three-fourth majority of the CMP as well as national ratification. This means that even if the amendment is passed, the compliance system becomes binding only on those countries that ratify it (Hovi et al. 2007, p. 446). With an enforcement system of this exceptional strength, it is likely that many states would have been hesitant to ratify, at least holding back to see what others would do before committing themselves. The anticipation of this possibility would also discourage others from ratifying the amendment out of fear of falling victim to these free riders. This would essentially turn participation in the compliance mechanism and the adherence to the commitments of the protocol more broadly, voluntary as the protocol would have no means of addressing cases of non-compliance of those that did not ratify. This would defy the logic of introducing an enforcement system in the first place. Making the enforcement system immediately binding for all, however, could only be achieved if it is adopted by a COP/CMP decision. Thus, an informal expansion was the only way to respond to the need for a strong enforcement system, and functional pressure can be confirmed as the main driving force.

The Solution: Expanding Force in Addition to Facilitation

The Kyoto compliance mechanism solves this complex of problems by integrating the managerial and enforcement model into one comprehensive and innovative system. It became clear that the managerial approach alone would not suffice to ensure compliance with the strict Kyoto targets. However, the possibility that non-compliance might result from a lack of capacity cannot be completely ignored. And, the argument that in the realm of environmental protection, prevention is more important than punishment still holds. This is also true for climate change, where CO2, once emitted, stays in the atmosphere for hundreds if not a thousand years (Inman 2008, p. 157). Thus, the managerial approach is not irrelevant or obsolete. This shows that the managerial approach and the enforcement model are not incompatible but can coexist. The two branches of the compliance committee each represent one of the two approaches to handling non-compliance. The bureau of the compliance committee will decide which approach is appropriate on a case-by-case basis by assigning it to the relevant branch.

The enforcement branch clearly follows, as the name implies, the enforcement approach. It comes into play as existing non-compliance is detected and "punishes" the non-complying state by imposing punitive or "hard" consequences. The nature of the enforcement branch is more judicial than political. It can not only be triggered by the parties but also by a neutral ERT. Once a party is found to be in non-compliance, there is little discretion for the application of consequences, and the CMP has only limited means to interfere, meaning states can expect to face consequences without much room for discussion. The consequences the enforcement branch applies are aimed at increasing the cost of non-compliance above the cost of compliance. Parties found to be in non-compliance will have to prepare a plan on how to restore compliance and submit it to the enforcement branch, which will monitor its implementation. This is a cumbersome process with regular meetings and progress reports. While this is still a relatively soft sanction, it has the legal effect of imposing a new obligation on the party in question. Until this plan is fully implemented, the party in question is suspended from the flexibility mechanisms. Such a suspension prevents the party from being credited with transactions in meeting its emission targets, and any tonnes transferred in violation of the suspension are not valid. This can have significant economic and political relevance and "should thus be considered a hard sanction" (Ulfstein and Werksman 2006, p. 56). During this time, the party in question will only be able to work toward its target by reducing emissions at home, which tends to be more costly than funding CDM projects or trading in the carbon market. Further, the party in question will lose the many business opportunities which arise through these mechanisms. The costs increase even more when targets are missed. A party failing their emission target at the end of a commitment period will have to make up the amount it fell short plus an additional 30 per cent of this amount in addition to their new commitment in the next commitment period. The application of this additional amount is automatic and not subject to negotiation. Thus, under the assumption that subsequent commitment periods would be adopted and the party in question plans to take part in these commitments, non-compliance would clearly be more expensive than making a serious effort.

A neutral cost–benefit analysis would conclude that complying with the agreement is the more rational decision, and Annex I parties should be motivated to stick to their commitments. The expectation of these additional costs alone likely deterred many from risking non-compliance and kept them from ever having to face the compliance committee. Since the enforcement branch only addresses cases of non-compliance of Annex I parties, it is unlikely that the managerial approach alone would have had the same effect. However, it is noteworthy that most (but not all) parties which faced the enforcement branch were Countries with Economies in Transition. It could be argued that these cases specifically could have been addressed by a managerial approach as well since the aftermath of tumultuous changes might still have played a role in them facing difficulties, although most of these cases took place around 2010/11.

The facilitative branch, on the other hand, applies the managerial approach of addressing non-compliance. It shares many characteristics with the Montreal compliance system, which became the standard model for MEAs. In comparison to the enforcement branch ("shall apply"), the procedures and mechanisms of the compliance procedure give the facilitative branch more discretion in applying consequences to its cases ("shall decide on", Decision 24/CP.7, Annex). These consequences are typical for the managerial approach and aim to lower the cost of compliance, in particular for developing countries. They include the provision of advice and facilitation of assistance, facilitation of financial and technical assistance, and formulation of recommendations to the party concerned. The facilitative branch does not make legally binding determinations of non-compliance (Ulfstein and Werksman 2006, p. 45). Further, the facilitative branch is tasked to provide an "early warning of potential non-compliance" to prevent environmental harm before it occurs. This would technically allow parties to address issues early on when they are still easier to fix and avoid the costly inconveniences of the enforcement branch. The facilitative branch, however, has not been able to live up to this task, which is most strikingly illustrated by the case of Canada.

The case of Canada finally calls into question the effectiveness of the Kyoto compliance committee as a whole. Essentially, Canada managed to avoid both branches. The facilitative branches early warning system was not able to pick up on arising problems in relation to Canada's emission target. And, Canada avoided the consequences of the enforcement branch by leaving the Kyoto Protocol. However, context is important. When they announced to leave in 2011, the demise of the Kyoto Protocol was already apparent, and the subsequent negotiations on the Paris Agreement were beginning, which Canada re-joined. Thus, what this case really illustrates is not a design flaw of the Kyoto compliance system as such but of the fact that its effectiveness depended on the continuous adoption of subsequent commitment periods.

Conclusion

This chapter has shown that the creation of the exceptionally strong Kyoto compliance committee was driven by functional pressure resulting from the specific nature of the Kyoto commitments and the flexibility mechanisms. While facilitative aspects and an early warning function remained relevant just as much as in most MEAs, a purely managerial approach could not have ensured the successful operation of the protocol. The managerial approach is focused on lowering the cost of compliance by channelling financial and technical assistance toward those parties that are struggling. Since, under the Kyoto Protocol, non-compliance was predominantly an issue of those that are the donors of assistance and not those eligible for receiving it, this approach alone would not have been effective. Non-compliance among the industrialised world needs to be addressed in a different way. What was needed instead was an enforcement mechanism able to impose "hard" sanctions. The anticipation of strict consequences increases the cost of non-compliance, thus manipulating the cost-benefit calculations of Annex I parties in such a way that it becomes more rational to comply. Since this applies to all parties involved, the danger of falling victim to a free-rider and losing competitiveness decreases as well. Further, the functionality of the popular flexibility mechanisms depended on a reliable compliance mechanism to ensure that offsets and allowances would take on a marketable value. This came at a significant price for Annex I parties in the form of a comparably very high loss of authority to the almost judicial enforcement branch.

All of this, however, would only work if the compliance mechanism would be applied to all parties at the same time. This means that the adoption through an amendment was essentially out of the question, although the relevant Article technically required it. The fact that an amendment was on the agenda, but no progress was ever made underlines the intricate nature of this problem. Adopting the needed enforcement system in the form of an amendment would have created a significant loophole, giving states the opportunity to avoid subjecting themselves to its exceptionally strict rules by postponing or avoiding to ratify. This would have defied the logic of providing confidence that everyone would make the effort. Thus, an informal expansion through a COP/CMP decision was the only feasible way to resolve this issue.

The Kyoto compliance committee was successful in the sense that all parties except for Canada achieved their emission targets in the first commitment period, and all questions of implementation addressed by the enforcement branch were successfully resolved. It was also successful to some extent in providing the right conditions for the carbon markets, at least until "hot air" credits resulting from the collapse of the Soviet Union caused a price decline. However, the enforcement branch never quite had the chance to live up to its full potential since its design heavily depended on the adoption of subsequent commitment periods. Further, the facilitative branch struggled to become relevant and ultimately failed to pick up on the looming non-compliance of Canada. These failures, however, need to be seen in the broader context of the Kyoto Protocol's demise and are not inherent design flaws.

The Kyoto model has since remained the only one of its kind, with the Paris Agreement returning to a more traditional, strictly managerial approach. This is not surprising. The Paris Agreement does not share one of the central aspects discussed in this chapter. This is the strict distinction between Annex I and non-Annex I parties in the application of legally binding targets. Under the Paris Agreement, all parties – industrialised and developing countries as well – set emission reduction targets for themselves. These targets are not legally binding. A strict enforcement approach simply does not fit with this more self-determined nature. The Paris Agreement differs significantly from both Montreal and Kyoto in this regard. As such, it requires, once again, a different approach. This is resolved through a facilitative compliance committee in combination with a particularly strong focus on monitoring and transparency. Whether this approach will enable a more effective international action on climate protection remains to be seen.

The Adaptation Fund Board

The Adaptation Fund is a particularly peculiar case of authority expansion in the area of implementation decisions. It is operated by the AFB, a body specifically created for this purpose under the authority of the Kyoto Protocol and now moved to the Paris Agreement. The Adaptation Fund and the AFB are explicitly not part of the financial mechanism of the UNFCCC, which is operated by the GEF. Further, in stark contrast to the donor-driven GEF, the AFB is composed of a majority of developing country parties. Long before the creation of the GCF (see section "The Creation of the Green Climate Fund"), the creation of the Adaptation Fund and the AFB marks the first step toward an internalisation of financial assistance away from the GEF and into the realm of influence of the climate MEA itself.

Remarkably, this curious constellation was negotiated after an earlier decision had essentially placed the authority over the adaptation fund in the hands of the GEF. This earlier decision was part of the Marrakesh Accords, a fundamental document including many specifications of the skeletal Kyoto rules, which many states had waited for before ratifying the Kyoto Protocol. Thus, the final outcome was certainly not on the state's minds when they made the decision to ratify the protocol.

What was the driving force behind this particularly unusual authority expansion? This chapter argues that the creation of the Adaptation Fund and the AFB under the Authority of the CMP indeed happened in the absence of functional pressure. Rather, it was driven by the exceptional force with which developing countries fought for more control over adaptation finance – an issue that was very dear to their hearts but, for industrialised countries, was merely another part of development aid. Two characteristics inherent in the Adaptation Fund played in their favour. First, the main resource of the Adaptation Fund is a share of proceeds from the CDM, meaning it does not rely on industrialised country parties' donations. Second, the UNFCCC does not entail the same membership as the Kyoto Protocol; most importantly, the USA – which unquestionably dominates the UNFCCC financial mechanism – is a member of the one but not the other. Further, the absence of the USA pushed the EU into a more proactive role. When the initially unsurmountable differences between the EU and the developing countries subsided in the context of a broader view on the future of the climate MEA at Bali, a deal on the Adaptation Fund could be struck.

The Case: Recipient-Driven Adaptation Aid

The Adaptation Fund is very different from the many funds of the UNFCCC. Its origin lies in one of the three flexibility mechanisms of the Kyoto Protocol, the CDM. The CDM allows a country with an emission reduction target under the Kyoto Protocol to earn certified emission reduction credits (CERs) by implementing an emission reduction project in a developing country without such a reduction target. These CERs can be either counted towards their emission targets or traded. The idea is to give industrialised countries flexibility in meeting their targets while stimulating sustainable development in the global South. Article 12 of the protocol sets down the specifics of this mechanism. Paragraph 8 states that:

> The Conference of the Parties serving as the Meeting of the Parties to this Protocol shall ensure that a share of the proceeds from certified project activities is used to (…) assist developing country Parties that are particularly vulnerable to the adverse effects of climate change to meet the costs of adaptation.

This paragraph would become the basis for the creation of the Adaptation Fund. The way toward its creation, however, would be long and rocky and was characterised by a deep division between developing and industrialised country parties.

The specifics of the Kyoto Protocol were carved out at COP 7 in Marrakesh in 2001. The so-called Marrakesh Accords include Decision 10/CP.7, which establishes the Adaptation Fund. This fund "shall be financed from the share of proceeds of the clean development mechanism project activities and other sources of funding" (para. 2). Its purpose is to finance concrete adaptation projects and programmes in developing country parties to the protocol. This marks a significant departure from previously existing adaptation funding under the convention, which had mainly been directed at financing enabling activities rather than the implementation of specific projects (Horstmann 2011, p. 1090).

A serious point of contention would soon become the operating entity of the Adaptation Fund. Paragraph 4 of Decision 10/CP.7 sets down that:

> (…) the adaptation fund shall be operated and managed by an entity entrusted with the operation of the financial mechanism of the Convention.

Long before the creation of the GCF in 2010, at this point in time, the financial mechanism of the convention was operated exclusively by the GEF. Thus, the GEF was implicitly chosen as the operating entity of the Adaptation Fund with this decision. Most industrialised country parties had withheld their ratification of the Kyoto Protocol until after Marrakesh, meaning they ratified under the assumption that this would be the case.

Things would, however, turn out very differently. After lengthy negotiations in the SBI at COP 13/CMP3 in Bali in 2007, it was decided in Decision 1/CMP.3 (paras. 3, 4) that:

> (…) the operating entity of the Adaptation Fund shall be the Adaptation Fund Board (…), the Adaptation Fund Board shall be established to supervise and manage the Adaptation Fund, under the authority and guidance of the Conference of the Parties serving as the meeting of the Parties to the Kyoto Protocol, and shall be fully accountable to the Conference of the Parties serving as the meeting of the Parties to the Kyoto protocol, which shall decide on its overall policies in line with relevant decisions.

The operation of the Adaptation Fund and thus the management of the share of proceeds resulting from the CDM is hereby placed with the newly created AFB, which explicitly operates under the CMP and not – as the financial mechanism of the convention – the GEF. The GEF would provide secretariat services only, and the World Bank would act as trustee (1/CMP.4, paras. 3–4). In addition to developing strategic priorities, policies and guidelines as well as specific operational policies and criteria-based principles and modalities for implementation, the AFB decides "on projects, including the allocation of funds". It is also responsible for the monetisation of the certified emission reductions issued by the Executive Board of the CDM.

The AFB is remarkable in its composition. Members and alternates will be chosen based on the appropriate technical, adaptation, and/or policy expertise but serve as government representatives. Developing countries hold a comfortable majority of 10 to 6 members, each holding one vote. While they cannot, on their own, block decisions where consensus fails, they only need to convince one other member of the board to achieve the required two-thirds majority (Decision 1/CMP.3, para. 12). This represents a sharp contrast to the donor-dominated GEF (see also sections "The Creation of the Interim Multilateral Fund" in Chapter 4 and "The Creation of the Green Climate Fund" in this chapter for a more in-depth discussion of the GEF's structures). As set down in Decision 1/CMP.3, paras. 6 and 7, the AFB consists of 16 members and 16 alternates representing parties to the Kyoto Protocol,

> (…) taking into account fair and balanced representation among these groups as follows:
>
> (a) Two representatives from each of the five United Nations regional groups;
> (b) One representative of the small island developing States;
> (c) One representative of the least developed country Parties;
> (d) Two other representatives from the Parties included in Annex I to the Convention (Annex I Parties);
> (e) Two other representatives from the Parties not included in Annex I to the Convention (non-Annex I Parties).

The UN regional groups are African States, Asia-Pacific, Eastern Europe, Latin America and the Caribbean, and Western European Countries and other States. All African states, all members of the Asia-Pacific group and all members of the Latin America Group are considered developing countries and negotiate under the Group of 77 in the context of the UN as well as the UNFCCC. Further, all SIDs and all least developed countries (LDCs) are developing countries. Annex I of the UNFCCC includes all industrialised countries that were members of the OECD in 1992, plus countries with economies in transition, making the group of non-Annex I parties mostly developing countries.

While this might seem, at first glance, like a big win for developing countries, the success of the fund has so far been limited. The consistent flow of income that was expected to reach the Adaptation Fund through the CDM stayed far below expectations when global carbon prices collapsed. Since 2010, the Adaptation Fund has committed US$ 923.5 million to projects and programmes to date, including 132 concrete projects (AFB/EFC.30/3/Rev.1). These are small numbers in relation to the UNFCCC financial mechanism. The GEF has provided more than 2 billion US$ in grant financing and mobilised close to 13 billion US$ from other sources for approximately 440 adaptation projects and provided at least 4.2 billion US$, and leveraged another 38.3 billion US$ for more than 1000 mitigation projects and programmes (GEF 2023a).

Further, with no obligation for industrialised country parties to donate, the Adaptation Fund struggles to attract pledges. History tells us that many such funds have been created only to be abandoned by donors (Ciplet et al. 2013, p. 63). The unusual composition of the AFB is likely among the reasons for why the Adaptation Fund fell victim to this. With developing countries in the majority, donors feel that they have lost control over how funds are distributed (Lebel et al. 2017, p. 233). Among all funds of the climate MEA, the lack of donor privilege makes the Adaptation Fund the least attractive (Graham and Serdaru 2020, p. 699). This clearly shows the implications the creation of the AFB had for the authority of member states. While the GEF's weighted voting structures would have provided a privileged position for industrialised country parties, those that finance the CDM projects and donate to the Adaptation Fund, the AFB privileges developing countries that are the recipients of the Adaptation Fund money and the CDM projects. Thus, as opposed to the donor-driven financial mechanism of the convention, this fund turned out to be the opposite, driven by recipients. Further, the Adaptation Fund and the AFB were explicitly placed under the CMP of the Kyoto Protocol as opposed to the COP of the UNFCCC. This includes only those parties that ratified the Kyoto Protocol – a significant detail, taking into account that the USA, as the largest potential donor, did not ratify it.

Another complication is the duplication of activities financed by the Kyoto Protocol's Adaptation Fund and the Special Climate Change Fund (SCCF) of the convention and the GEF (Dessai 2003, p. 301). It has also been argued that since the creation of the GCF in 2010, this new fund has "sucked up all the oxygen in the system" (TANGO International 2015, p. 54).

Still, although its ultimate success remains limited, the Adaptation Fund and the AFB are a striking case of authority expansion of the climate MEA and even reversed a decision that had already been made. Thus, it clearly warrants further inquiry.

Cultivated Pressure: Fighting for Survival in the Absence of Functional Need

Despite being established in the Marrakesh Accords in 2001, negotiations on the governance of the Adaptation Fund only took up momentum at the 24th meeting of the SBI in May 2006. This meeting turned out to be a major failure, caused by insurmountable differences between developing and industrialised country parties. The main issue at stake was the operating entity of the Adaptation Fund. Industrialised country parties, the EU in particular, continued to support the GEF. Japan, for example, argued for the GEF for reasons of "minimisation of transaction costs and complementarity with other funds". The G77 and China, on the other hand, especially LDCs, entirely rejected it. The Philippines, for example, argued that there was great risk that projects under the Adaptation Fund would attain the same conditionality as other GEF funds and added that "the further problem with the GEF is that developing countries, and especially LDCs, are not properly represented, the representation through constituencies is not truly democratic" (Grasso 2011, p. 370). Many developing countries, in particular the most vulnerable, consider the GEF and its management procedures to be awkward and extremely inefficient. They have continuously called for the GEF to be more responsive to the guidance of the UNFCCC, to facilitate access to existing funding for adaptation and to give greater consideration to adaptation priorities (Grasso 2011, p. 368). Further, a central issue is that GEF funding is conditional on creating "global benefits" – which essentially excludes adaptation projects (GEF 2023b). While this condition has been eased through the creation of the SCCF under the GEF, which explicitly funds adaptation projects, this fact further undermined the trust of developing countries in the GEF to handle adaptation issues in their best interest.

Still, it seems unlikely that the climate MEA as a whole could have failed because of this issue. If industrialised country parties had insisted on the GEF solution, developing countries could have blocked the CDM or left the Kyoto Protocol altogether. Both would not have meant the end of the protocol. Unlike other cases discussed in previous chapters (e.g., the creation of the Interim MLF of the Montreal Protocol in section "The Creation of the Interim Multilateral Fund" in Chapter 4), the participation of developing countries in the Kyoto Protocol was not necessary for the achievement of its aims. Emission reduction targets were only set for industrialised countries. There explicitly were no new commitments for developing countries in the Kyoto Protocol (see Article 10). The main area of action for developing countries under the protocol would be to host CDM projects. Developing countries themselves had been split on the CDM. In the Kyoto negotiations, Brazil had led the pro-CDM side together with the USA, while China and India showed continued resistance (Stokes et al. 2016, p. 25; Bettelli et al. 1997, p. 15). The CDM, together with the protocol's other flexibility mechanisms, had been largely a US initiative, and at this point, they had long left the agreement (Dessler and Parson 2019, p. 29). Further, the EU had argued, alongside developing countries, that domestic action should be the main focus to achieve emission targets. The Marrakesh Accords provide that "the use of the mechanisms shall be supplemental to domestic actions" in meeting commitments (Decision 15/CP.7). The EU had insisted on this supplementarity and also proposed a ceiling for the use of flexibility mechanisms, which was, however fiercely opposed by Japan, USA, Canada, Australia, Norway, and New Zealand as well as the Russian Federation and thus never became a reality (Gusbin et al. 1999, p. 833).

It is somewhat tricky to assess if Annex I parties could have achieved their targets without the CDM. Still, the available data give some indications. Industrialised country parties accepted a higher overall target in exchange for the flexibility mechanisms (from 5 to 6 per cent). But it seems likely that they could still have achieved this even without the CDM. At the end of the Kyoto Protocol's first commitment period, only 4.4 per cent of the EUs emission reductions were achieved through CERs, 2,2 per cent for Japan, 0.8 per cent for Australia, 4.0 per cent for New Zealand, and 0.0 per cent for Russia. These numbers look even smaller in light of the fact that the EU exceeded its reduction target by 3 per cent, Australia by 4 per cent, Japan by almost 10 per cent, Norway by 10 per cent, and New Zealand by 43 per cent (UNFCCC 2019). Norway explicitly bought CERs only in addition to domestic action through a programme aimed at buying from projects which were at risk of suspension because of low market prices (Norwegian Ministry of Climate and Environment 2014). What had made compliance easier, though, was the large amount of so-called "hot air" sold through the international carbon market. These were credits sold by Russia and Ukraine, which resulted not from actual emission reductions but were leftovers from the collapse of the Soviet economy between the protocol's 1990 baseline year and its 1997 adoption. It cannot clearly be established how much "hot air" was sold, but published data suggest about 2.2 GtCO2e per year, which is slightly less than the parties' collective overcompliance. This indicates that the role of these "hot air" credits was, ultimately, limited from an overarching perspective (Dessler and Parson 2019, pp. 30–33).

Thus, while a good number of states were in favour of the CDM, all taken together, it seems unlikely that its potential collapse could have been a large-scale deal breaker. Industrialised country parties likely could have achieved their targets even without the CDM, and developing country participation was not absolutely necessary for the Kyoto Protocol to survive. What is more, the CDM had already begun its operation 1 year prior, with the first projects being approved in 2005. Thus, it seems unlikely that developing countries could have had any true intention of blocking the CDMs operation.

Further, even if this conflict could have meant the demise of the Kyoto Protocol, this would not have meant the end of the climate MEA as a whole. When the Kyoto Protocol eventually did collapse for a plethora of other reasons (see also section "The Creation of the Green Climate Fund"), the MEA did not come to a halt but renewed itself with the creation of the Paris Agreement. Indeed, when discussing the institutional structures of the Adaptation Fund, this was on the negotiator's minds already. At the Bali conference, where the creation of the AFB was officially decided at CMP level, the Bali Action Plan was adopted – a roadmap for negotiations on the future of climate cooperation beyond the first commitment period of the Kyoto Protocol.

Taken together, it can be confidently stated that in this case, no functional pressure was present to trigger the creation of the Adaptation Fund under the authority of the developing-country-dominated AFB under the CMP. If no action had been taken, the Adaptation Fund would have been created under the GEF – as had been implicitly decided in Marrakesh. While this certainly was not the preferred option for developing countries and could have caused some distress, this chapter has shown that it is unlikely that the issue could have posed a serious threat to the climate MEA as a whole. Indeed, in Marrakesh, developing countries had technically already accepted the GEF solution. Thus, this peculiar authority expansion came about driven by a group of actors – the developing countries (G77 plus China) – that had changed their minds since Marrakesh and now had a strong interest in this particular expansion. The remainder of this chapter will investigate how they were able to accomplish this expansion against a legally binding decision that had already been made by employing the structures of the MEA in their favour and making use of favourable conditions to build a powerful coalition.

The Solution: The Protocol versus the GEF

A dramatic shift had taken place in global climate change negotiations at this point in time. The issue of adaptation to the adverse impacts of climate change has taken centre stage, reaching the top of the agenda alongside the mitigation of climate change. At the core of the politics surrounding adaptation funding is the reality that the poorer developing countries have contributed almost nothing to causing climate change, yet they are experiencing the impacts first and worst. While many of the G77 early on wanted more focus on adaptation, for Annex I parties, this would be an acknowledgement of responsibility and liability. Further, mitigation projects in Non-Annex I countries still benefit Annex I countries by reducing global GHG concentrations. However, the far-reaching benefits of adaptation projects are less obvious (Ciplet et al. 2013, p. 51). A World Bank study that has systematically tried to separate the additional costs due to climate change above and beyond existing development aid and adaptation deficits arrived at a figure of US$ 70–100 billion per year in developing countries alone (Narain et al. 2011, p. 1003). For the developing world, climate change is a matter of sheer survival, and the financing of concrete adaptation actions to prevent or reduce climate impacts has become a crucial form of defence. For the industrialised world, however, climate change is mostly an environmental problem, albeit a very serious one, and their emotional temperature consequently is much milder. They tend to see adaptation financing as additional development assistance which should be mainstreamed into existing development activities (Grasso 2011, p. 362). Thus, there is a strong asymmetry in the importance of this issue among the two opposing groups. The matter of adaptation funding would be one where developing countries fight vigorously to gain control while industrialised parties would remain merely avoidant.

Two unique aspects of the Adaptation Fund gave developing countries leverage to realise their preference against that of industrialised country parties. First, it is funded by an international revenue-generating source, the CDM, and thus is not tied to the national budgets of industrialised countries. At SBI 25, the Philippines argued that "The governance structure should reflect the main sources of funding for the AF, that is, the share of the proceeds from CERs activities under the CDM" (Grasso 2011, p. 372). Technically, the 2 per cent levy on the CDM that funds the Adaptation Fund belongs to the host nation where investments are bought, not the purchaser. Therefore, this fund is essentially paid for by developing countries themselves (Ciplet et al. 2013, p. 63). The Adaptation Fund is merely the mechanism through which they share the funds among themselves and, in particular, with those that have very limited mitigation capabilities and can, therefore, not host CDM projects on their own (FCCC/SBI/2006/11, Annex II). This makes the fund particularly important for LDCs which need adaptation funding the most and, at the same time, have no chance to voice their needs in the GEF (Grasso 2011, p. 368). This means that the Adaptation Fund is not dependent on donors, a crucial difference from all other funding mechanisms under the climate MEA and other similar funds. Thus, the donor side was significantly weaker in these negotiations than they usually are in similar situations.

Second, the CDM and the Adaptation Fund explicitly operate not under the UNFCCC as all other parts of the financial mechanism but under the Kyoto Protocol, a treaty which does not include the same membership, most notably not the USA (Ciplet et al. 2013, p. 62). The fact that the Adaptation Fund belongs to the Kyoto Protocol and not the convention implies that it is outside the influence of those that did not ratify the protocol. Developing countries continuously emphasised this fact to avoid a predisposition in favour of the existing financial mechanism. Since the USA was not part of the Kyoto Protocol, it would not sponsor any CDM projects and thus not create any revenue for the fund. In the GEF, however, the USA had a particularly strong voice. Thus, integrating the Adaptation Fund into the GEF would, as argued by developing countries, represent an objective misrepresentation. Further, the exclusion of the USA represented an opportunity for other big players, such as the EU, to assume a more proactive role in the negotiations (Grasso 2011, p. 363).

While strongly in favour of the GEF solution at SBI 24, the EU had changed its mind in preparation for Bali. Indeed, the Bali conference was characterised significantly by the emergence of a coalition between developing countries and the EU. A key factor for the success of the negotiations on the Adaptation Fund was the pre-Bali declaration by the EU that they would accept whatever model for the design of the Adaptation Fund the G77 endorsed (Müller 2008, p. 1). The purpose of this coalition was likely to gain the goodwill of the developing countries for the adoption of the Bali Action Plan on the future of climate cooperation beyond the Kyoto Protocol. A central part of this plan is the consideration of including developing countries in mitigation actions "in a measurable, reportable and verifiable manner" (Decision 1/CP.13, para. 1. (b) (ii)). The unprecedented willingness of developing countries to take up a more active role in the fight against climate change shown at Bali effectively annihilated the main excuse of the US administration for not acting on climate change, namely the unwillingness of developing countries to make a contribution. As such, the conciliatory position of the EU did not only succeed in soothing tensions between developing and industrialised country parties but also acted as a bridge leading the USA back into the fold (Ott et al. 2008, p. 93). This broader view of the situation at COP7 can add to the understanding of why the industrialised country party side eventually succumbed to the immense pressure developing countries exerted on the issue of the governance of the Adaptation Fund. While it seems unlikely that the EU's undertaking as a whole could have been endangered by the discussion on the Adaptation Fund and the AFB at this point, it surely was part of a larger package which allowed for a more positive environment to unfold and launch more fruitful negotiations about the climate MEAs future. It has also been argued that the Adaptation Fund was a *quid pro quo* for developing countries' acceptance of the voluntary nature of donations to this fund and the two other funds created simultaneously under the GEF, the SCCF, and the LDCs Fund (Dessai 2003, p. 298).

Conclusion

This chapter has shown that the creation of the Adaptation Fund and the AFB under the authority of the Kyoto Protocol was a particularly peculiar authority expansion of the climate MEA in the area of implementation decisions. What makes this case even more remarkable is that this expansion happened in direct opposition to an earlier decision and, as this chapter has shown, even in the absence of functional pressure.

The creation of the Adaptation Fund and the AFB clearly followed an actor-driven logic. By then, adaptation had turned into a central issue for developing country parties. For many, it was a matter of survival. Thus, they were highly motivated to achieve control over this source of funding. As a result, they achieved an authority expansion which went far beyond the mere functionality of the MEA. Meanwhile, industrialised country parties did not share the same passion and – although they did not give up easily, eventually they accepted defeat. Moreover, the EU managed to turn their loss on this issue into a strategic benefit for the future of the climate MEA as a whole.

The group of developing countries inside the MEA had a strong interest in achieving this expansion, and they made the existing institutional logics and structures work in their favour, in particular the inherent logic of the CDM and its donated share of proceeds. As put into words by the Philippines, it was easy to argue that the fund should allow developing countries significant influence, since this would more closely reflect the source of the money. This is in sharp contrast to the funds under the GEF, which are financed through industrialised country donations and thus follow a logic that favours an argument for the donor-driven structures of the GEF. Further, emphasising that the Adaptation Fund explicitly belonged to the protocol and not the convention gave developing countries another point of leverage. Giving the Adaptation Fund into the hands of the US-dominated GEF when the USA was not a member of the Kyoto Protocol was hard to justify. And, with the usually all-dominating USA not a part of the negotiations, the donor side was significantly weakened, and those that were more sympathetic to the developing countries' cause took a more proactive role. When the EU switched sides and favoured a coalition with developing countries at Bali, the Adaptation Fund under the AFB became a reality. Taken together, these points made the developing countries strong enough to change a decision that had already been made.

However, this lack of functional pressure might also be part of the reason for the Adaptation Fund's limited success. With no clear need for the Adaptation Fund to stay strong and many alternatives where they would enjoy more privilege, donors have largely steered away.

The Creation of the Green Climate Fund

The creation of the GCF in 2010 is a very recent case of authority expansion under the climate MEA, and it is one that not only fundamentally transformed the structures of international climate protection but also has consequences for our understanding of MEAs and their role in the global landscape. The GCF is a new kind of funding institution in the emerging field of climate finance. Its structure is innovative in its equal representation of developed and developing countries on its board, its pursuit of equal mitigation and adaptation financing, and its mandate to engage directly with the private sector. The most striking aspect, however, is the fact that the GCF is exceptionally independent, with legal capacity and judicial personality of its own – created by decision of the UNFCCC and without formal amendment or new founding treaty.

The creation of the GCF represents an exceptionally extensive authority expansion that follows into the footsteps of the creation of the MLF of the Montreal Protocol. And indeed, the conflict about its structures were very similar. Since the GCF does not include specific legally binding financial obligations, it is not an expansion of financial implementation assistance as such, although the promise of additional support was an important aspect. Instead, since it moves the authority over funding decisions away from the donor-driven GEF and closer to the control of the COP, it is a case of expansion in the area of implementation decisions. Where the financial mechanism of the UNFCCC was originally operated by the GEF alone – with its close affiliation to the World Bank, its weighted voting, and its donor-driven approach – the GCF would now serve as a second operating entity which does not allow the same donor privileges. Thus, the creation of the GCF significantly transformed the power structures inside the UNFCCC financial mechanism. It took a part of the control over financial decisions away from donors and into the hands of recipients. But, this came not without a cost to recipients as well.

This chapter argues that the creation of the GCF was the watershed moment that made the adoption of the Paris Agreement – more precisely, the adoption of a new agreement with emission targets for all member states – possible. There was surely a feeling that not just the future of the climate MEA but of the planet as a whole was at stake. Due to emissions growing in the Global South through development and the threat of carbon leakage undermining the success of climate protection on a global scale, there was strong functional pressure to create a new agreement that would call developing countries to account and play their part in protecting the climate. The creation of the GCF essentially became part of a package deal which gave developing country parties new financial assistance they could equally control in exchange for their acceptance of their own emission targets. Under the Paris Agreement, both financial obligations, as well as emission targets, are soft and based on a self-determined, bottom-up approach. While the effectiveness of this can surely be criticised, it likely made this extensive deal more acceptable to both sides. Further, similar to the creation of the MLF of the Montreal Protocol, the establishment of the fund through a formal amendment would have created a complicated transitory period, which would delay the creation of the fund as a whole as well as allow states to avoid financial responsibilities by delaying their ratification. Interestingly, in contrast to the Montreal Fund, the creation of the GCF was not later formalised through an amendment, but its creation rests purely on a COP decision.

The Case: A New Operating Entity

The GCF first appeared in the Copenhagen Accord, a short document of a very particular nature. Negotiated in parallel to the official COP 15 discussions by only a small number of parties, it does not represent an official COP decision but was merely "noted" by the COP (Decision 2/CP.15). In the Copenhagen Accord, the states involved set down that:

> 10. We decide that the Copenhagen Green Climate Fund shall be established as an operating entity of the financial mechanism of the Convention to support projects, programmes, policies and other activities in developing countries related to mitigation including REDD-plus, adaptation, capacity-building, technology development and transfer.

Although it asserts that the Accord will be "operational immediately", fully operationalising its terms required further action, in particular the establishment of the envisioned fund. And, the terms of the Copenhagen Accord presumed that this would be carried out by the COP, although it had not officially adopted this document (Bodansky 2010b, p. 238).

One year later, the GCF appeared on the official COP agenda at its 16th Meeting in 2010 in Cancun. A COP decision was adopted, which now officially created the GCF and designated it as a new operating entity for the UNFCCC financial mechanism. In paragraph 102 of the Cancun Agreements (Decision 1/CP.16), the COP:

> Decides to establish a Green Climate Fund, to be designated as an operating entity of the financial mechanism of the Convention under Article 11, with arrangements to be concluded between the Conference of the Parties and the Green Climate Fund to ensure that it is accountable to and functions under the guidance of the Conference of the Parties, to support projects, programmes, policies and other activities in developing country Parties using thematic funding windows.

Notably, there was no formal amendment involved which would have required individual ratification, not even after the establishment of the fund as it had been the case with the Montreal fund. The GCF's legal basis is solely a COP decision. This is particularly striking since Article 11 of the UNFCC, which establishes the MEA's financial mechanism, clearly states that "Its operation shall be entrusted to one or more existing international entities" (emphasis added). The GCF, however, was an entirely new creation. While it is accountable to the COP, the GCF possesses juridical personality of its own and has "such legal capacity as is necessary for the exercise of its functions and the protection of its interests". Further, the fund is supported by a secretariat which is "fully independent" (GCF Governing Instrument, Decision 1/CP.16, para.108), making the GCF an exceptionally independent entity – founded by the COP of the UNFCCC and not by the individual member states.

The GCF is governed by a Board which has "full responsibility for funding decisions", giving it complete control over how the financial assistance flowing through the fund will be spent. The Board has 24 members from an equal number of industrialised and developing countries. Representation from developing country parties includes representatives of relevant UN regional groupings and representatives from SIDs and LDCs. Decisions of the Board were originally taken by consensus. In July 2018, however, the GCF board experienced serious gridlock, which led to the agreement that if all attempts to reach consensus are exhausted, decisions can be made with a qualified majority of two-thirds. This procedure was used for the first time in relation to a Chinese project at the 24th board meeting in November 2019. Japan and the USA voiced objections, but in the resulting vote, the project was approved by a qualified majority of 19 members, with two against and one abstention, over-ruling two of the largest donors (Kalinowski 2020, p. 6)

All developing country parties to the convention are eligible to receive funding from the GCF to finance agreed incremental costs for activities to enable and support enhanced action on adaptation, mitigation, technology development and transfer (including carbon capture and storage), capacity-building, and the preparation of reports. The fund supports developing countries in pursuing project-based and programmatic approaches in accordance with climate change strategies and plans, such as low-emission development strategies or plans, nationally appropriate mitigation actions, national adaptation plans of action, national adaptation plans, and other related activities. The GCF Board decided to aim for a 50:50 balance between mitigation and adaptation activities over time (GCF.B.06/06).

The GCF's central objective and guiding principle is to "contribute to the achievement of the ultimate objective of the United Nations Framework Convention on Climate Change" (GCF Governing Instrument 2011, I. (2)). More specifically, its goal is to raise new and additional funds of US$ 100 billion per year by 2020 from a variety of sources, including public, private, bilateral, and multilateral and alternative sources to finance climate change-related projects in developing countries (FCCC/CP/2009/11/Add.1, FCCC/KP/CMP/2010/12/Add.1). The GCF's Governing Instrument enables the fund to accept pledges from developed country parties to the UNFCCC as well as public, non-public, and alternative sources. Such sources include, among others, countries not party to the UNFCCC, entities, and foundations. Contributions from parties to the UNFCCC and other sovereign entities may be made in the form of grants, capital, or loans. In its initial resource mobilisation phase, the GCF has raised US$ 10.3 billion in pledges from 49 countries/regions/cities, including 9 developing countries (Chile, Colombia, Indonesia, Mexico, Mongolia, Panama, Peru, Republic of Korea, and Viet Nam) and as of January 31, 2022, 34 contributors have pledged US$ 10 billion for the first formal replenishment (GCF 2020).

The creation of the GCF is an interesting case because of its exceptional independence. While accountable to the COP, it has its own independent secretariat, juridical personality, and legal capacity. It does not create specific, binding financial obligations, however, its creation expands the authority of the COP to control how donations will be disbursed. Where it had been the donor-driven GEF alone operating the financial mechanism of the UNFCCC before, there is now a second entity working under the authority of the COP alongside it, which manages a significant amount of the newly mobilised financial flows. And, this entity is not an existing one as required by Article 11 but an entirely new one, created for this specific purpose by the UNFCCC itself.

The decisions on which projects will be funded are made by the GCF board with its equal representation of donor and recipient countries. This is a powerful step away from the donor-driven weighted voting structures of the GEF. It decreases the control of donors over how their money would be used and increases the influence of those that will receive it. Notably, unlike in the case of the MLF of the Montreal Protocol, there was no formal amendment and ratification process to formalise the creation of the GCF. Not everyone was happy with this development. Indeed, the GCF was one of the main reasons cited by Donald Trump for leaving the Paris Agreement. Among his more erratic condemnations, like claiming that the (notably non-binding) US pledge of 1 billion US$ included “funds raided out of America’s budget for the war against terrorism” (World Resource Institute 2017), US representatives did call for a more donor-driven approach in the GCF (Kalinowski 2020, p. 3).

Functional Pressure: The Need to End Carbon Leakage

One thing was clear when the parties to the UNFCCC came together in Copenhagen: A more global way of addressing climate change had to be found, which would replace the Kyoto Protocol’s focus on emission targets for industrialised nations only (Bodansky 2010b, pp. 232–234). The Kyoto Protocol had run into serious trouble at this point. The most striking issue was the decision by the USA not to ratify. The 2000 election had brought an abrupt change to US climate efforts. In the negotiations of the Kyoto Protocol, the Clinton administration had been inconsistent but ultimately reached a satisfying compromise. The newly elected Bush administration, however, expressed clear hostility to the protocol and to international climate cooperation in general. Thus, in March 2001, President Bush announced that the USA would not ratify the Kyoto Protocol. The reasons he cited were scientific uncertainty, the high costs of emission reductions, and the lack of obligations for developing countries (Dessler and Parson 2019, p. 30). With the USA as one of the largest emitters worldwide not included, the efficiency of the protocol was seriously undermined.

However, there was a larger issue behind this incident. The fact that the Kyoto Protocol did not include any emission reduction obligations for developing countries was about to seriously undermine any effort to protect the climate. Although countries with binding targets collectively reduced their emissions during the period covered by the Kyoto Protocol, global emissions continued to increase (Nielsen et al. 2021, p. 149). As developing countries began to industrialise rapidly, many became large emitters of GHGs. While emissions were reduced by 23 per cent in Europe, they increased by 95 per cent in Asia from 1990 to 2010 (OECD 2015). Indeed, China became the world's largest emitter of CO2 in 2007, surpassing the USA (Ivanova 2017, p. 19). This shows that the US argument is valid to some extent: "nothing developed countries do will matter if the large emerging economies are not held accountable for their rapidly growing emissions" (Clémençon 2016, p. 5). However, the real issue is more complex. Some amount of the growing emissions in developing countries were indeed not completely made up of their own, but they were caused by carbon "leakage" – the re-location of emissions from the territory of industrialised countries with emission reduction targets to that of developing countries where no such targets were in place. This leads to a situation in which emissions are not reduced but merely moved. The only certain way to eliminate leakage and make effective climate protection possible is to convince all regions of the world to participate in at least some kind of emission targets, no matter how weak they might be at the start (Dessler and Parson 2019, p. 150).

It became clear that, in order to achieve the aims of the climate MEA, a new agreement had to be created, which would include reduction targets for as many states as possible worldwide. Of course, though, developing countries were reluctant to accept that they would now have to play their part in reducing GHG emissions as well, and they would not agree to it without getting anything in return.

Leading up to Copenhagen, the Philippines, for the G-77/China, recalled convention Article 4.7 (ENB 2009, p. 9), which makes developing countries' commitments conditional on developed country assistance:

> The extent to which developing country Parties will effectively implement their commitments under the Convention will depend on the effective

implementation by developed country Parties of their commitments under the Convention related to financial resources (…).

Finance became a "make or break" element of the negotiation deal in Copenhagen. Barbados, for AOSIS, called for significant public funding and asked industrialised countries how much money they would be "putting on the table". Financial support, together with technology transfer, soon became "carrots to persuade developing countries to accept the sticks of a new climate regime in which all countries would bear the costs" (Oh 2020, p. 2). It became clear that a new financial arrangement had to be found which would satisfy developing countries so that they would be willing to accept taking on their own emission targets.

The general need for new and additional finance was quickly accepted. But, similar to the case of the MLF of the Montreal Protocol, it was not only the amount which would become central to the negotiations but also the question of how it would be decided how the money would be spent. Once again, a situation arose in which developing countries were not willing to accept costly reduction targets without a strong promise of financial support, more specifically, support which would not be dominated by donor countries (see section "The Creation of the Interim Multilateral Fund" in Chapter 4). Developing countries generally supported a new financial mechanism, including a MLF under the authority and guidance of the COP with balanced geographical representation, similar to the MLF of the Montreal Protocol. Industrialised countries, on the other hand, called for the use of existing institutions. Once again, the GEF would become the centre of attention – or, more precisely, contention.

Different options were on the table (FCCC/TP/2008/7, paras. 436–440). Some proposals called for making use of and improving existing institutions, in particular, of course, the GEF, to avoid a proliferation of institutions and funds, while others preferred the establishment of new institutional arrangements. Developing countries were vocal in proposing the creation of various new funds under the authority of the COP. The G77 and China called for the establishment of a financial mechanism which would operate under the supreme authority and guidance of, and be fully accountable to, the COP. This mechanism would have an equitable and geographically balanced representation of all parties within a transparent and efficient system of governance. Similarly, Mexico proposed the establishment of a "World Climate Change Fund" operating under the COP with an inclusive and transparent governance system. Some parties explicitly proposed that the financial mechanism "should be modelled on the Multilateral Fund for the Implementation of the Montreal Protocol" (para. 440).

It soon became clear that any type of arrangement commanded by the GEF alone (or any other World Bank affiliate for that matter) was unlikely to satisfy developing countries and would not be a viable option to resolve the functional tensions at stake. The GEF predates the UNFCCC and was established as a pilot programme at the World Bank in 1991, making it subject to the World Bank's weighted voting system, which favours large donor countries. Because of this and because of its focus on evaluation and cost-effectiveness, donors have traditionally been quite fond of it (Graham and Thompson 2015, p. 128), and quite the opposite has been true for the recipient side. Following the adoption of the UNFCCC, the GEF gained independence from the World Bank in 1994 and adopted a constituency voting system, which, however, remained similar to the World Bank's (Graham and Serdaru 2020, p. 688). The 176 member states of the GEF are represented through 32 constituencies, and decisions require a double majority of members and of 60 per cent of contributions. While this does provide some amount of greater formal influence for developing countries, they continue to criticise the GEF for its close association with the World Bank and perceive it as a donor-dominated institution with weak democratic legitimacy. Indeed, the number of states in a constituency varies widely. For example, the USA, China, France, Germany, Italy, Canada, and Japan are each a constituency unto themselves and hold their own seat on the Council, while Bangladesh, Bhutan, India, Maldives, Nepal, and Sri Lanka share a single seat. Those in multistate constituencies question the democratic legitimacy of a system where the loyalties of representatives (e.g. to the nation versus the larger constituency) are in question, leaving some to feel they are represented in name only. The significance of formal voting rules (and, for that matter, changes in voting rules) is modest, however, given that a formal vote has never occurred and perceptions that donors drive decision-making still persist (Graham and Thompson 2015, p. 128). As such, further reforming of the GEF's voting rules would be unlikely to convince developing countries that the GEF would move away from its donor-driven logic. The perception held by developing countries that the GEF is dominated by wealthy states provided an important impetus for the establishment of the GCF, which incorporates stronger representation and voting rights for developing states (Graham and Serdaru 2020, p. 690).

In this context, the South Centre's Global Governance Programme for Development remarked in a submission to a UNFCCC working group (FCCC/AWGLCA/2008/MISC.3/Add.1) that:

> The use of the joint World Bank-UNDP-UNEP Global Environment Facility (GEF) as currently the sole operating entity for the Convention's financial mechanism has been fraught with many implementation challenges which developing countries have also long critiqued, as well as with challenges with respect to its compliance with the requirements of the financial mechanism under Art. 11 of the Convention. The COP, in its decisions, has been consistent in recognising that the Convention does not limit the choice of operating entities for the financial mechanism to only the GEF.

Indeed, paragraph 1 of Article 11 on the financial mechanism specifies that "Its operation shall be entrusted to one or more existing international entities". Paragraph 2 further stipulates that "The financial mechanism shall have an equitable and balanced representation of all parties within a transparent system of governance". Both are formulated in a legally binding way ("shall"). Indeed, the GEF does not fulfil the requirements of paragraph 2.

It follows that there was not only functional pressure to expand the financial mechanism but even more so to move it away from the GEF as its only operating entity. Further, existing alternatives were rare. At the time, no international entity existed which realistically could have taken on this task, and which also fulfilled the conditions of equity, balance, and transparency. Thus, creating an entirely new operating entity which could take on the task and fulfil the required conditions was the only viable option.

Further, this could not have been achieved through the adoption of an amendment. An amendment would only become binding for those that ratified it and only enter into force on the ninetieth day after the date of ratification by at least three-fourths of the parties to the convention (Article 15). Similar to the Montreal Protocol's MLF, there would be a transitory period of significant length in which some members would have ratified the amendment and others would not. During this period, only those that ratified could be expected to contribute to the fund. If larger donors delayed or entirely avoided ratification, this would leave the fund with no significant resources to disburse and essentially turn it into an empty shell. This would defy its function of providing developing countries with a quid pro quo for their acceptance of reduction targets. Under Montreal, this issue was solved by the creation of the Interim Fund to cover the in-between period. In this case, however, a similar arrangement would have taken time and trust in each other and the future, which parties did not have at this point. The first commitment period of the Kyoto Protocol would end in 2012, and the outlook for the adoption of a second one was not promising. Negotiations on the future of climate protection had broken down, and no new agreement had been adopted as planned in Copenhagen in 2009. Global climate protection was effectively in the process of coming to a halt. It is not only unlikely that an amendment of this magnitude would have found much support under these conditions. It would also put back the timeline even further, leaving the future of the agreement unclear for even longer until it would enter into force. The immediate creation of the GCF through a collectively binding COP decision was the only way to get everyone back at the negotiating table in a timely manner and pave the way for the eventual adoption of the Paris Agreement in 2015. Thus, it can be concluded that functional pressure was clearly at the root of this informal expansion of authority of implementation decisions over financial decisions.

The Solution: Expanding COP-Control

The constellation which was finally agreed on can be seen as a compromise between donor and recipient interests: The financial mechanism would be entrusted to two operating entities, which would each represent the preferences of the opposing groups. One part would remain within the donor-driven GEF, and a second part would be entrusted to the GCF, a new institution accountable to the COP and free from the World Bank's legacy.

The GCF's Governing Instrument was hailed as a "progressive, forward-looking document" (Carlarne 2012, pp. 20–21). It reflects a world in which the Global South has gained a stronger voice. The GCF fulfils the criteria set down in Article 11 of equitable and balanced representation and transparency. It has learned from the critiques associated with the World Bank. The GCF operates with parity between donors and recipient countries in decision-making. Unlike in the World Bank, decisions are not be dominated by donors but made based on the communication between donors and recipients as formally equal partners. When it comes to decisions on projects, Kalinowski (2020, p. 7) even argues that they are recipient-driven. Recipients do not just have parity in the boards but they also have superior knowledge of the projects they propose. This gives them the upper hand in project approvals, while donors can merely block proposals that they consider inadequate. The GCF is also more transparent than other institutions, its board meetings are even streamed live online. Further, the GCF follows a polycentric, multistakeholder approach that integrates the regional level, civil society organisation as well as the private sector. Thus, it is driven neither by powerful individual member states nor by a central hierarchy. Its design, its decisions, and the implementation of policies and projects arise out of the interaction of different stakeholders at different levels.

The addition of the GCF as a second operating entity to the financial mechanism would remedy developing countries' dissatisfaction with and distrust of the GEF and create a support structure that would give them the necessary confidence to join a new agreement which would require them to adopt emission targets. Although the Copenhagen Accord did not bring the desired new long-term agreement at first, it paved the way for the eventual adoption of the Paris Agreement only a few years later and commit all parties to submit GHG reduction targets in their NDCs. As Oh (2020, p. 560) states:

> There is no exaggeration in asserting that the establishment of the Green Climate Fund of the Financial Mechanism at COP16 constituted a watershed, opening the door for the eventual conclusion of the Paris Agreement, which governs post-2020 climate governance.

It is safe to say that without the creation of the GCF, the adoption of the Paris Agreement would have been unlikely. While the adequacy of the Paris Agreement and its bottom-up approach in the face of the devastating dangers of climate change can be questioned, it is a crucial step on the way towards achieving the aims of the climate MEA. The Kyoto Protocol had clearly failed and would not be able to achieve any significant advances in climate protection or even continue to run in a meaningful way. A new agreement was necessary, which would include emission reduction targets for all countries. Without developing country participation, in particular China but also India and Brazil, any efforts to reduce emissions and manage climate change would come to nothing – not only because of their own growth but also because of carbon leakage from the industrialised world. Any type of target, no matter how small, is a serious step towards reducing leakage. Further, if developing countries would not participate, the participation of large, industrialised parties, the USA in particular, would be in question as well. The EU 27 together emitted 3.15Gt CO2e in 2019, much less than the USA (5.77Gt) or China (12.06Gt) alone (Climate Watch Data 2019). Even if a smaller number of particularly motivated industrialised states, like those that accepted emission targets under the second commitment period of the Kyoto Protocol (which unfortunately only entered into force on the day it expired), would have continued the effort on their own, this would not have been enough to reduce emissions to a level that could prevent devastating climate change.

The GCFs institutional form is modelled after the Montreal Protocol's MLF, which was also created as an alternative to the GEF. The GCF and the MLF are similar in many ways but also differ in several aspects. The most striking difference in the context of this chapter is the fact that the creation of the MLF was formalised through a later amendment which required individual ratification. Although the creation of the fund was, at that point, already a done deal, this does provide the MLF with a more stable legal basis. This is particularly striking since the MLF is slightly less independent than the GCF. While the MLF explicitly operates "under the authority of the Parties who shall decide on its overall policies" (Decision II/8, para. 4 of the Montreal MOP), the GCF, created only by COP decision (or, looking at the Copenhagen Accord originating in a document with an even more doubtful legal nature) merely operates under the "guidance" of its COP. A reason for this might be the difference in financial obligations involved with the two funds. The MLF includes specific financial obligations for donor states, which might have induced the parties to ensure some stricter control over managing their contributions. The GCF relies on voluntary pledges and private sector involvement, which might allow for or even require less direct oversight by the COP. Further, these differences reflect the different general approaches of the two entities. The specific financial obligations of the MLF intertwine with the specific common emission reduction obligations of the Montreal Protocol, while the GCF pledges fit with the bottom-up approach of NDCs, which would enter the Paris Agreement.

The GCF did, however, turn out not to be quite as equal and independent as it might have been expected. While no official decision has ever explicitly allowed earmarking, a number of contribution agreements between the GCF Trust Fund and donors ended up including "targets", which were subsequently identified as earmarks. For example, the USA, Canada, and the UK target their contributions toward the Private Sector Facility. Australia even notes in its contribution arrangement that "the Australian Government has requested that our contribution should facilitate, including through the GCF's Private Sector Facility, private-sector-led economic growth in the IndoPacific region" (GCF 2015). It has been argued that these permissive earmark rules are, in their essence, a design substitute for the weighted voting of the GEF. They can act as a backstop for donor control where developing countries otherwise control governing-body decision-making (Graham and Serdaru 2020, p. 672). The possibility to earmark contributions weakens the authority of the GCF board. It effectively cedes some control over the distribution of resources to individual donors (Graham 2017, p. 366). Permitting these earmarks across windows makes the GCF's ability to meet its 50:50 distribution between mitigation and adaptation contingent on the donor's individual funding decisions. Although negotiations have demonstrated that developing countries increasingly view these permissive earmarking rules as violating the principle of equity, they might have no choice but to accept them. There is suggestive evidence that if wealthy states fail to achieve asymmetric control in one place, some will shift their funds elsewhere (Graham and Serdaru 2020, p. 695).

Conclusion

This chapter has shown that the creation of the GCF as the second operating entity of the UNFCCC financial mechanism was driven by functional spillover. Carbon leakage resulting from the Kyoto Protocol introducing reduction targets for industrialised countries only caused pressure to innovate the agreement in such a way that it would mobilise reduction efforts more globally. Similar to the case of the Montreal fund, this was a case in which developing countries could successfully leverage the need for their participation to re-model the financial mechanism of the MEA in their favour. The desperate need for their participation gave the usually weak developing countries an unusually powerful position to make demands. They were quick to point out Article 4.7 of the UNFCCC treaty, which makes their implementation of commitments dependent on the provision of financial resources. If they were to accept targets, they wanted more extensive financial support in return. Similar to the case of the Montreal fund, they did not only demand additional financial support but also a funding arrangement which would grant them equal standing in the decision-making process. Thus, any type of arrangement under the donor-driven GEF would be inacceptable. As such, further functional pressure was created, which led to the creation of the GCF under the guidance of the COP as an entirely new second operating entity of the financial mechanism. This made the financial mechanism acceptable enough for developing countries to pave the way for their acceptance of emission targets. To ensure that the GCF would actually take shape in a meaningful way and a timely manner, it had to be created through a COP decision that relied on the lengthy process of ratification, which would give states the option to delay their acceptance. Essentially, the GCF had to be created to make the adoption of the Paris Agreement possible.

Finally, although the GCF was successful in making the adoption of the Paris Agreement possible, it remains questionable if the innovations of the GCF will be sufficient to empower the Global South in a broader sense. Criticism of the GCF is voiced broadly and loudly. While (recipient) country ownership and civil society involvements were criticised by the Trump administration, private sector involvements are criticised by civil society groups. Further, scholars criticise "greenwashing business as usual (BAU) projects and subsidies to private companies" and "the commodification and financialisation of the environment as such" (Kalinowski 2020, p. 4). Further, the environmental integrity of the GCF has been called into question when the board refused an explicit ban on fossil fuel projects (Upadhyay 2015) and when it accredited banks like the Japanese Sumitomo Mitsui Banking Corporation, the HSBC, Deutsche Bank, and Credit Agricole, which are all heavily involved in financing high-level fossil fuel projects (Lo 2021).

References

Bettelli, Paola; Carpenter, Chad; Davenport, Deborah; Doran, Peter; Wise, Steve (1997): Report of the Third Conference of the Parties to the United Nations Framework Convention on Climate Change. 1–11 December 1997. *Earth Negotiations Bulletin* 12 (76), pp. 1–16.

Bodansky, Daniel (2010b): The Copenhagen Climate Change Conference: A Postmortem. *The American Journal of International Law* 104 (2), pp. 230–240. DOI: 10.5305/amerjintelaw.104.2.0230

Brunnée, Jutta (2003): The Kyoto Protocol. A Testing Ground for Compliance Theories? *Heidelberg Journal of International Law* 3 (63), pp. 225–280.

Carlarne, Cinnamon P. (2012): Rethinking a Failing Framework. Adaptation and Institutional Rebirth for the Global Climate Change Regime. *Georgetown International Environmental Law Review* 25 (1), pp. 1–50.

Ciplet, David; Roberts, J. Timmons; Khan, Mizan (2013): The Politics of International Climate Adaptation Funding: Justice and Divisions in the Greenhouse. *Global Environmental Politics* 13 (1), pp. 49–68. DOI: 10.1162/GLEP_a_00153

Clémençon, Raymond (2016): The Two Sides of the Paris Climate Agreement. *The Journal of Environment & Development* 25 (1), pp. 3–24. DOI: 10.1177/1070496516631362

Climate Action Tracker (2023): *Temperatures. 2100 Warming Projections*. Available online at https://climateactiontracker.org/global/temperatures/, checked on 08/07/23.

Climate Watch Data (2019): *Historical GHG Emissions*. Available online at www.climatewatchdata.org/ghg-emissions?end_year=2019®ions=TOP&source=CAIT&start_year=1990, checked on 09/07/23.

Dagnet, Yaminde; Northrop, Eliza (2017): Facilitating Implementation and Promoting Compliance (Article 15). In Daniel Klein, María Pía Carazo, Meinhard Doelle, Jane Bulmer, Andrew Higham (Eds.): *The Paris Agreement on Climate Change. Analysis and Commentary*. Oxford: Oxford University Press, pp. 338–351.

Dessai, Suraje (2003): The Special Climate Change Fund: Origins and Prioritisation Assessment. *Climate Policy* 3 (3), pp. 295–302. DOI: 10.3763/cpol.2003.0334

Dessler, Andrew E.; Parson, Edward A. (2019): *The Science and Politics of Global Climate Change*. Cambridge: Cambridge University Press.

Doelle, Meinhard (2021): Non-Compliance Procedures. In Lavanya Rajamani, Jacqueline Peel (Eds.): *The Oxford Handbook of International Environmental Law*. Second Edition. Oxford: Oxford University Press, pp. 972–987.

ENB (2009): *Earth Negotiation Bulletin: AWG-LCA 5 and AWG-KP 7 Highlights*: Monday 30 March 2009. Available online at https://enb.iisd.org/events/bonn-climate-change-talks-marchapril-2009/daily-report-30-march-2009, checked on 2/10/2020.

GCF (2015): *Contribution Arrangement with Australia (IRM)*. Available online at www.greenclimate.fund/document/contribution-arrangement-australia-irm, checked on 09/07/23.

GCF (2020): *Resource Mobilisation*. Available online at www.greenclimate.fund/about/resource-mobilisation, checked on 09/07/23.

GEF (2023a): *Climate Change Mitigation*. Available online at www.thegef.org/what-we-do/topics/climate-change-mitigation, checked on 1/8/2023.

GEF (2023b): *How Projects Work*. Available online at www.thegef.org/projects-operations/how-projects-work, checked on 2/8/2023.

Graham, Erin R. (2017): The Institutional Design of Funding Rules at International Organizations: Explaining the Transformation in Financing the United Nations. *European Journal of International Relations* 23 (2), pp. 365–390. DOI: 10.1177/1354066116648755

Graham, Erin R.; Serdaru, Alexandria (2020): Power, Control, and the Logic of Substitution in Institutional Design: The Case of International Climate Finance. *International Organization* 74 (4), pp. 671–706. DOI: 10.1017/S0020818320000181

Graham, Erin R.; Thompson, Alexander (2015): Efficient Orchestration? The Global Environment Facility in the Governance of Climate Adaptation. In Erin R. Graham, Alexander Thompson, Kenneth W. Abbott, Philipp Genschel, Duncan Snidal, Bernhard Zangl (Eds.): *Efficient Orchestration? International Organizations as Orchestrators*: Cambridge: Cambridge University Press, pp. 114–138.

Grasso, Marco (2011): The Role of Justice in the North–South Conflict in Climate Change: The Case of Negotiations on the Adaptation Fund. *International Environmental Agreements* 11 (4), pp. 361–377. DOI: 10.1007/s10784-010-9145-3

Gusbin, Dominique; Klaassen, Ger; Kouvaritakis, Nikos (1999): Costs of a Ceiling on Kyoto Flexibility. *Energy Policy* 27 (14), pp. 833–844. DOI: 10.1016/S0301-4215(99)00075-0

Harvey, Fiona (2015): Paris Climate Change Agreement: The World's Greatest Diplomatic Success. *The Guardian*. Available online at www.theguardian.com/environment/2015/dec/13/paris-climate-deal-cop-diplomacy-developing-united-nations, checked on 08/07/23.

Hoffmann, Matthew J. (2013): Global Climate Change. In Robert Falkner (Ed.): *The Handbook of Global Climate and Environment Policy*. Chichester, West Sussex, Malden, MA: Wiley, pp. 3–18.

Höhne, Niklas; Elzen, Michel den; Rogelj, Joeri; Metz, Bert; Fransen, Taryn; Kuramochi, Takeshi et al. (2020): Emissions: World Has Four Times the Work or One-Third of the Time. *Nature* 579 (7797), pp. 25–28. DOI: 10.1038/d41586-020-00571-x

Horstmann, Britta (2011): Operationalizing the Adaptation Fund: Challenges in Allocating Funds to the Vulnerable. *Climate Policy* 11 (4), pp. 1086–1096. DOI: 10.1080/14693062.2011.579392

Hovi, Jon; Bretteville Froyn, Camilla; Bang, Guri (2007): Enforcing the Kyoto Protocol: Can Punitive Consequences Restore Compliance? *Review of International Studies* 33 (3), pp. 435–449. DOI: 10.1017/S0260210507007590

Inman, Mason (2008): Carbon is Forever. *Nature Climate Change* 1 (812), pp. 156–158. DOI: 10.1038/climate.2008.122

IPCC (1992): *Climate Change. The 1990 and 1992 IPCC Assessments, IPCC First Assessment Report Overview and Policymaker Summaries and 1992 IPPC Supplement*. Geneva: IPCC.

IPCC (2014): *Climate Change 2014: Synthesis Report. Contribution of Working Groups I, II and III to the Fifth Assessment Report of the Intergovernmental Panel on Climate Change*. Geneva, Switzerland: IPCC.

Ivanova, Maria (2017): Politics, Economics, and Society. In Daniel Klein, María Pía Carazo, Meinhard Doelle, Jane Bulmer, Andrew Higham (Eds.): *The Paris Agreement on Climate Change. Analysis and Commentary*. Oxford: Oxford University Press.

Kalinowski, Thomas (2020): Institutional Innovations and Their Challenges in the Green Climate Fund: Country Ownership, Civil Society Participation and Private Sector Engagement. *Sustainability* 12 (21). DOI: 10.3390/su12218827

Lebel, Louis; Kallayanamitra, Chalisa; Salamanca, Albert (2017): The Governance of Adaptation Financing: Pursuing Legitimacy at Multiple Levels. *International Journal of Global Warming* 11 (2), pp. 226–245. DOI: 10.1504/IJGW.2017.10001237

Lefeber, René; Oberthür, Sebastian (2012): Key Features of the Kyoto Protocol's Compliance System. In Jutta Brunnée, Meinhard Doelle, Lavanya Rajamani (Eds.): *Promoting Compliance in an Evolving Climate Regime*. Cambridge: Cambridge University Press, pp. 77–101.

Lo, Joe (2021): *Coal-Backing Japanese Bank Bids for Green Climate Fund Partnership*. Available online at https://climatechangenews.com/2021/03/15/coal-backing-japanese-bank-bids-green-climate-fund-partnership/, checked on 09/07/23.

Maslin, Mark (2004): *Global Warming. A Very Short Introduction*. Oxford: Oxford University Press.

Müller, Benito (2008): *Bali 2007: On the Road Again! Impressions from the Thirteenth Climate Change Conference*. Oxford: Oxford Institute for Energy Studies.

Narain, Urvashi; Margulis, Sergio; Essam, Timothy (2011): Estimating Costs of Adaptation to Climate Change. *Climate Policy* 11 (3), pp. 1001–1019. DOI: 10.1080/14693062.2011.582387

Nielsen, Tobias; Baumert, Nicolai; Kander, Astrid; Jiborn, Magnus; Kulionis, Viktoras (2021): The Risk of Carbon Leakage in Global Climate Agreements. *International Affairs* 21 (2), pp. 147–163. DOI: 10.1007/s10784-020-09507-2

Nordhaus, William D.; Boyer, Joseph (2000): *Warming the World. Economic Models of Global Warming*. Cambridge, MA: MIT Press.

Norwegian Ministry of Climate and Environment (2014): *Carbon Credits*. Available online at www.regjeringen.no/en/topics/climate-and-environment/climate/innsiktsartikler-klima/klimakvoter/id2076655/, checked on 2/8/2023.

Obergassel, W.; Arens, C.; Beuermann, C.; Hermwille, L.; Kreibich, N.; Ott, H. E.; Spitzner, M. (2020): COP25 in Search of Lost Time for Action: An Assessment of the Madrid Climate Conference. *Carbon & Climate Law Review* 14 (1), pp. 3–17. DOI: 10.21552/cclr/2020/1/4

OECD (2015): *COP21: Climate Change in Figures*. Available online at www.oecd.org/environment/cop21-climate-change-in-figures.htm

Oh, Chaewoon (2020): Contestations over the Financial Linkages between the UNFCCC's Technology and Financial Mechanism. Using the Lens of Institutional Interaction. *International Environmental Agreements* 77 (2017), p. 49. DOI: 10.1007/s10784-020-09474-8

Ott, Hermann; Sterk, Wolfgang; Watanabe, Rie (2008): The Bali Roadmap: New Horizons for Global Climate Policy. *Climate Policy* 8 (1), pp. 91–95. DOI: 10.3763/cpol.2007.0510

Rajamani, Lavanya; Werksman, Jacob (2021): Climate Change. In Lavanya Rajamani, Jacqueline Peel (Eds.): *The Oxford Handbook of International Environmental Law*. Second Edition. Oxford: Oxford University Press, pp. 492–511.

Stokes, Leah C.; Giang, Amanda; Selin, Noelle E. (2016): Splitting the South: China and India's Divergence in International Environmental Negotiations. *Global Environmental Politics* 16 (4), pp. 12–31. DOI: 10.1162/GLEP_a_00378

Sunstein, Cass R. (2007): Of Montreal and Kyoto. A Tale of Two Protocols. *Harvard Environmental Law Review* 31 (1), pp. 1–65.

TANGO International (2015): *First Phase Independent Evaluation of the Adaptation Fund. With Assistance of in Association with the Overseas Development Institute*. Washington, DC: World Bank.

Ulfstein, Geir; Werksman, Jacob (2006): The Kyoto Compliance System. Towards Hard Enforcement. In Olav Schram Stokke, Jon Hovi, Geir Ulfstein (Eds.): *Implementing the Climate Regime. International Compliance*. Sterling: Earthscan, pp. 39–62.

UNEP (2019): *Emissions Gap Report*. Nairobi: UNEP.

UNFCCC (2019): *Compilation and Accounting Data – First Commitment Period of the Kyoto Protocol*. Available online at https://di.unfccc.int/flex_cad, checked on 2/8/2023.

UNFCCC (2023): *Compliance under the Kyoto Protocol*. Available online at https://unfccc.int/process/the-kyoto-protocol/compliance-under-the-kyoto-protocol, checked on 10/21/2023.

Upadhyay, Anand (2015): *Green Climate Fund Can Be Spent To Subsidise Dirty Coal*. Available online at https://cleantechnica.com/2015/04/07/green-climate-fund-can-spent-subsidise-dirty-coal/, checked on 09/07/23.

Weart, Spencer R. (2008): *The Discovery of Global Warming*. Cambridge, MA: Harvard University Press.

Werksman, Jacob (1998): Compliance and the Kyoto Protocol. Building a Backbone into a Flexible Regime. *Yearbook of International Environmental Law* 9, pp. 48–101.

Werksman, Jacob (2006): The Negotiation of a Kyoto Compliance System. In Olav Schram Stokke, Jon Hovi, Geir Ulfstein (Eds.): *Implementing the Climate Regime. International Compliance*. Sterling: Earthscan, pp. 17–38.

World Resource Institute (2017): *Fact-Checking Trump on Climate Finance*. Available online at www.wri.org/insights/fact-checking-trump-climate-finance, checked on 09/07/23.

6 Conclusion

It has been shown that Multilateral Environmental Agreements (MEAs) are a peculiar hybrid form of cooperation. While they are commonly understood as mere treaties – as empty shells structuring state interaction, the analysis undertaken here has shown that this view severely underestimates their true nature as separate entities on the global landscape and their potential to expand their authority over time, creating new obligations for member states without their explicit consent. This final chapter will review the steps undertaken and summarise the results achieved on an individual as well as a generalisable level. This book has developed a new and innovative theoretical framework to analyse this phenomenon and its specific driving forces by combining recent approaches to the study of authority with the Neofunctionalist concept of spill-over. Six case studies uncovered the unique authority relationship that emerges between MEAs and their members and the driving forces leading to significant shifts. Informal authority expansions in MEAs happen where gaps and ambiguities in the existing institutional framework become relevant and the ratification process necessary for formal treaty amendments would pose a significant obstacle to resolving the issue. These expansions are predominantly, but not exclusively, driven by a functional need to address additional issues collectively. Interestingly, the cases investigated here further all share their roots in conflicts arising between developing and industrialised country parties. These results significantly advance our understanding of informal institutional dynamics in environmental politics and beyond, and will be of interest not only to those investigating MEAs specifically but also those interested in new forms of authority on the international level and other forms of informal governance and institutional adaptation.

Review of the Steps Undertaken

Chapter 1 has reviewed established perspectives on MEAs in the fields of International Relations (IR) and International Law. Scholars of International Law recognise MEAs as a distinct form of cooperation that functions almost identically to full-fledged International Organisations (IOs, Churchill and

DOI: 10.4324/9781003597681-7

Ulfstein 2000; Brunnée 2002; Wiersema 2009). The fact that MEAs are specifically established with a more informal, ad hoc nature should not be a reason to deny them legal personality. Since states can influence the Conferences of the Parties (COPs) of MEAs only by acting through them, they need to be considered as more than the sum of their members and essentially possess "a will of their own" (Churchill and Ulfstein 2000, p. 633). In sharp contrast, the field of IR continues to have trouble grasping MEAs as a distinct phenomenon and tends to conceptualise them as mere treaties, functioning as arenas for state interaction (Keohane 1989; Krasner 1982; Bauer 2006; Barnett and Finnemore 1999). They remain intervening variables with no existence separate from the states they are composed of. Because of the focus on the rational–legal authority of the state or large-scale bureaucracies of IOs prevalent in these approaches, the capabilities of more ad hoc and informal organisations like MEAs remain underestimated.

More recent approaches to the study of authority in global governance have opened up a path to establish an understanding of MEAs from an IR perspective that more closely relates to that of the field of law. Chapter 2 has traced the emergence of these approaches. Where the concept of authority used to be exclusively bound to the rational–legal legitimacy of the state, newer approaches recognise the concept of *relational authority* as "a social contract in which a governor provides a political order of value to a community in exchange for compliance by the governed with the rules necessary to produce that order" (Lake 2010, p. 587). Applied to the international landscape, this means that international institutions have authority when states recognise them as having authority, specifically by accepting the decisions they make as legally binding on their own actions (Cooper et al. 2008, p. 505). The concept of relational authority provides a starting point to grasp the true nature of MEAs more accurately. Chapter 2 has shown that it can be applied to the study of MEAs, establishing an understanding of this special type of institution as distinct entities capable of influencing the international landscape in their own right through the performance of different governance functions.

Based on this new understanding, the phenomenon of MEA expansion moves to centre stage. Once we begin to apply an understanding of authority as a dynamic, two-dimensional relationship, it becomes apparent that significant shifts are possible without formal delegation. Chapter 3 has established a new and innovative theoretical framework to analyse these informal MEA authority expansions. The basis is provided by the concept of *conversion*, an informal version of the classic agency slack. This describes a situation in which rules remain formally the same, but by making use of gaps and ambiguities in the contract, they can be implemented in new ways (Heldt and Schmidtke 2017, p. 54; Mahoney and Thelen 2009, p. 16). Transferred to the study of MEAs, this describes situations in which the rules of the original treaty remain formally untouched, but COPs take on new functions through their own interpretations of these rules through their decisions.

Chapter 3 has further derived two hypotheses on the specific driving forces which make these expansions possible based in neofunctionalism as one of the most central theories of institutional expansion (Niemann 2006; Haas 1958; Lindberg 1963; Tranholm-Mikkelsen 1991). First, a functional logic based on a need to address an additional issue on the collective level to keep the MEA functioning was suspected. Since MEAs are specifically designed to be able to respond flexibly to newly emerging challenges, this is expected to be the main driving force of informal MEA authority expansion. Second, an actor-driven logic loosely based on the concept of cultivated spill-over has been suspected, in which actors inside the MEA strategically employ the MEA's skeletal rules to create opportunities for expansion that create benefits beyond the desired order of value provided by a functioning MEA.

Chapters 4 and 5 have analysed six cases of informal MEA authority expansion to identify the main driving force and illustrate the phenomenon more broadly. As one of the most long-standing MEAs that is also regarded as the most successful to date, the Montreal Protocol provided the ideal starting point. Chapter 4 has focused on three cases under the Montreal Protocol on Substances that Deplete the Ozone Layer, the creation of the Interim Multilateral Fund, which served as the blueprint for many MEAs that came after it, the decision to classify developing countries on a case-by-case basis through Meeting of the Parties (MOP) decision, and the creation of the compliance mechanism which was the first to follow a managerial approach. All three of these cases were primarily driven by a functional need to address additional issues on the collective level.

The section "The Creation of the Interim Multilateral Fund" in Chapter 4 has shown that the creation of the Interim Multilateral Fund was driven by strong functional pressure resulting from a mismatch between strict obligations (which were soon set to tighten even further) and the lack of commitment to financial assistance for building the necessary capacities in the developing world. Developing country participation was essential for the Montreal Protocol to effectively protect the ozone layer. If they continued and possibly even expanded their use of ozone-depleting substances, all efforts under the protocol would be undermined. However, many developing countries had put their participation on hold for fear that they would compromise their development by adhering to the strict obligations. Thus, the provision of financial support became necessary. This strong need put developing country parties in a favourable position to make demands about how this support would be managed. Traditionally opposed to the World Bank, they did not accept any type of structure under its donor-driven weighted voting. And thus, the creation of a new fund under the authority of the one-country–one-vote principles of the MOP became the only viable solution, even though most industrialised country parties had strictly opposed this option. The creation of this fund was later formalised through a treaty amendment. However, all structures and obligations came to life before this amendment entered into force based on a MOP decision. This was necessary because an amendment required formal ratification by the individual member states.

This created a significant transitory period in which some would have ratified the amendment while others had not. Would the amendment enter into force before major donors had ratified it, others would have to finance a much larger share of the overall cost than originally planned. Since this was an unacceptable risk, an interim arrangement that would become immediately binding on all parties was necessary, and this could only be provided by MOP decision. This created significant financial obligations for industrialised country parties, which had not been foreseen when the protocol was first adopted, and the largest donors were unsuccessful in securing privilege over how this money would be used.

The section "Defining Developing Countries" in Chapter 4 demonstrated that the decision to classify developing countries under the protocol on a case-by-case basis through MOP decisions without objective criteria was driven by functional pressure spilling over from tightened control measures and the creation of the Multilateral Fund. Since the relevance of the classification grew significantly with these changes, a flexible approach was needed to ensure that the fragile willingness of developing countries to participate would not be upset. All existing lists and criteria that were considered would have categorised some states in a way that they were unlikely to accept. Further, since the classification is tied closely to the obligations under the protocol, the issue necessarily had to be addressed on the collective level since a common understanding among all was needed for the protocol to function. This put the decision on which parties would be eligible for support from the Multilateral Fund and which would have to pay into it into the hands of the collective and out the hands of each individual state.

The section "Compliance Management" in Chapter 4 illustrated that the establishment of the Montreal Protocol's new compliance management procedure was driven by a functional need to ensure compliance with the obligations of the Montreal Protocol, and therefore the achievement of the Protocol's aims under the specific nature of environmental issues. The collective nature of the environment as a common good made a mechanism necessary that could address compliance from a collective point of view. Neither relying on the existing dispute settlement procedure nor the establishment of stricter enforcement rules could have fulfilled this purpose. Dispute settlement is focused on conflicts between individual parties and is usually triggered by an individually injured party. In environmental protection, however, it is not an individual party that is injured by non-compliance but the environment as a collective good. Further, environmental harm that is produced by non-compliance can hardly ever be fully restored. Thus, for a MEA to be effective, in most cases, it will have to focus on the prevention of harm over the punishment after the fact. The collective nature of this issue also meant that a compliance mechanism needed to apply to all parties collectively and at the same time, which cannot be achieved through an amendment requiring ratification but only through a MOP decision. The creation of this new type of system put the management of compliance into the hands of the collective of the parties to the protocol.

Chapter 5 has investigated three cases of authority expansion under the UN Framework Convention on Climate Change (UNFCCC), the Kyoto Protocol, and the Paris Agreement. With climate change being the most pressing environmental issue of our time, it is now under constant observation by the public eye and is thus central to any study of global environmental politics. Further, its' so far limited success and complicated history stand in sharp contrast to the Montreal Protocol's success, thus providing a different perspective on environmental governance. Three cases of authority expansion in the realm of climate protection were investigated, including the creation of the Kyoto enforcement branch as a compliance mechanism with exceptional strength, the creation of the adaptation fund board based on developing countries' fight for survival in a changing climate, and the creation of the Green Climate Fund (GCF) to pave the way for the adoption of the Paris Agreement. The first and the last of these cases were driven by a functional need for collective solutions. The creation of the Adaptation Fund Board with its specific recipient-driven structure, however, took place in the absence of functional pressure and was the result of developing country parties strategically pushing for more control over adaptation finance.

The section "The Kyoto Enforcement Branch" in Chapter 5 carved out how the creation of the exceptionally strong Kyoto compliance mechanism was driven by a functional need rooted in the specific nature of the Kyoto commitments. Since the Kyoto Protocol includes specific emission targets only for industrialised country parties, a purely managerial approach rooted in the facilitation of access to assistance analogous to the Montreal system would not have been efficient since they themselves are the donors of financial aid and drivers of technological innovation. In this context, a purely managerial approach would not have been sufficient to provide enough confidence among industrialised country parties that everyone would make the effort so they would make it themselves. Confidence was further essential to ensure that carbon credits would take on a marketable value so that the popular market mechanisms would become functional. Similar to the Montreal mechanism, however, the system needed to be adopted in the form of decision since an amendment would have created a significant loophole, giving states the opportunity to avoid subjecting themselves to its rules by postponing or avoiding ratification. This system extensively expanded the collective control over the compliance of industrialised country parties and is even mandated to impose sanctions on those that miss their targets.

The section "The Adaptation Fund Board" in Chapter 5 represents the only one of the six cases analysed here which happened in the absence of functional pressure. The creation of the Adaptation Fund Board with its recipient-driven structure under the authority of the CMP was driven by the exceptional force with which developing countries fought for more control over adaptation finance – an issue that was very dear to their hearts but for industrialised countries was merely another part of development aid. What played in their favour was the fact that the Adaptation Fund is financed by a share of proceeds from the Clean Development Mechanism, meaning it does not rely on industrialised country parties' donations. Further, it is explicitly established under the Kyoto Protocol and not the UNFCCC. Since both treaties do not entail the exact same membership (in particular, the USA not being a party to the protocol), there was good reason not to integrate the Adaptation Fund into the UNFCCC's financial mechanism but to place it under a new committee. Finally, with the USA moved out of focus of the discussion, the European Union was able to attain a more proactive role and broker a deal with developing country parties. Placing the Adaptation Fund under the Adaptation Fund Board and the CMP did not only preclude the USA from influencing this particular channel of climate finance. The fact that its structures privilege developing countries means that the usually dominating donor side, in this case, had to accept a minor role in influencing how this money would be used.

The section "The Creation of the Green Climate Fund" in Chapter 5 established that the creation of the GCF was driven by a functional need to make a new agreement possible, which would include commitments for all parties globally. Growing emissions in developing countries and extensive carbon leakage severely undermined the efforts made under the Kyoto Protocol. Thus, developing countries needed to be convinced to take on emission targets of their own. Similar to the creation of the Multilateral Fund of the Montreal Protocol, the GCF (and its creation through decision over amendment) became what made global participation possible. Once again, developing countries strictly opposed any type of solution involving donor privilege, which excluded the World Bank and also the GEF as the existing operating entity of the climate financial mechanism. Creating the GCF with its equal representation structures as the second operating entity of the financial mechanism ensured that the developing world would have its own voice that would be heard while the donors lost some of their privilege over climate finance.

Beyond Individual Case Study Results

Comparing the results of each individual case beyond the boundaries of their respective MEAs provides interesting insights into the phenomenon of MEA authority expansion more broadly. Figure C.1 illustrates how cases can be clustered according to the results obtained.

In five of the six investigated cases, functional pressure has been identified as the central driving force of authority expansion. These expansions seem to happen predominantly early on as the skeletal rules of the original treaty are fleshed out through subsequent COP negotiations and slow down as the MEA operates continuously over a longer period of time. As has been shown, they can be catalysed by outside events like the collapse of the Soviet Union, which are not themselves a trigger of expansion but might cause existing tensions to become more relevant and call for a more immediate resolution. All five could be traced back to gaps or ambiguities in the existing regulatory framework, which needed to be closed or clarified in order to keep the MEA functional. These five cases can be classified into two groups of issues, which

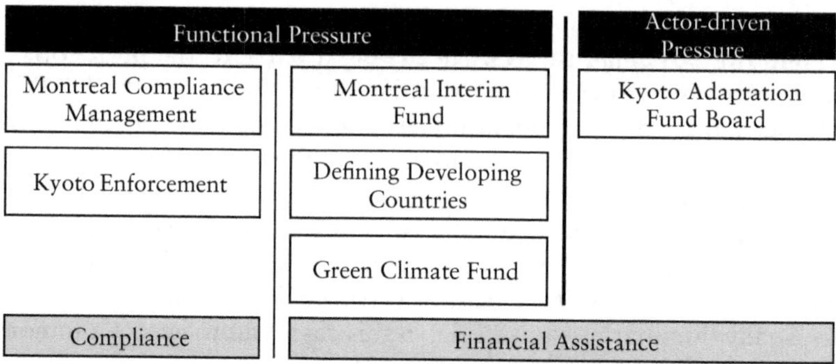

Figure C.1 Summary of case study results.

both occurred in similar ways under each of the two MEAs analysed here. These two areas are compliance and financial assistance.

First, both the Montreal Protocol and the Kyoto Protocol went through the process of developing compliance mechanisms which took on unexpected designs. Although the resulting expansions are diametral opposites of each other, one being the first to follow a managerial approach and the other being the only strict enforcement mechanism in the realm of global environmental governance to date, the functional structure of the underlying need that led there are, in principle, two sides of the same coin. The collective nature of the atmosphere means that both cases deal not with injured individual parties but with an injured collective good. This means that traditional dispute settlement will not be functional to solve these problems. Instead, non-compliance needs to be addressed from a collective perspective and applied collectively. Where Montreal and Kyoto differ is in which parties were subject to specific targets. Where developing country parties had to achieve their own targets, the focus needed to be on support; meanwhile, where only those that are the providers of support have to adhere to targets, a different approach is needed, and enforcement became central.

Second, both the Montreal Protocol and the climate MEA encountered a situation in which developing country participation became necessary to achieve the aims of the MEAs but developing countries would not or could not accept strict environmental targets without the promise of extensive financial implementation assistance. This lack of sufficient support to give developing countries enough confidence to join created a functional need for an expansion of implementation assistance. Being traditionally extremely critical of the World Bank and the GEF and strictly opposed to any type of donor privilege, any type of arrangement under their direct control would not be acceptable. Thus, a functional need for funding mechanisms specific to the MEAs emerged, which led to the creation of the Multilateral Fund and the GCF, respectively. The case of defining developing countries under the Montreal Protocol fits into this group as well since it was driven by the need to clarify eligibility for the fund.

Functional pressure seems to be the predominant but not the only driving force of informal MEA authority expansion. One of the cases analysed here, the case of the Adaptation Fund Board, took place in the absence of functional need. A combination of different factors has been identified that allowed developing countries to achieve this expansion. Interestingly, this expansion is also closely connected to financial implementation assistance and evolves around similar issues as the functionally driven cases discussed in this context, as it was focused on the strong wish to exert more control over how financial support would be disbursed and avoiding the donor-driven structures of the World Bank and the GEF. Further research is necessary to analyse this second driving force more in-depth for a more robust understanding of how it operates.

A further aspect that has been discovered along the way is that all six cases share a core factor: They are all rooted in circumstances resulting from the divide between industrialised and developing country parties that is woven into the very essence of both the Montreal Protocol and the climate MEA. This is more obvious in the cases addressing financial assistance, where the central line of conflict falls between the donor and recipient side. But, it was also relevant in the cases addressing non-compliance. Under the Montreal Protocol, the expectation was that non-compliance would be predominantly a problem of the developing world. The compliance mechanism's managerial approach is built around the central notion of bringing developing country parties back on track and not deter their participation by imposing strict consequences. In the case of the Kyoto enforcement branch on the other hand, only industrialised country parties had to adhere to specific emission targets. Thus, the compliance mechanism had to be built, instead, on disciplining industrialised countries, which required a very different approach. In the two MEAs analysed here, the developing and the industrialised world face widely differing situations in the light of their responsibility for causing the relevant problems and their capability to contribute to resolving them – and thus also in their obligations under the MEAs. Interestingly, authority expansions seem to happen most prominently in areas where these differences matter most – finance and compliance.

Based on these results, several conclusions can be drawn beyond the individual institutions for MEAs as a specific type of organisation more generally:

First, it has been conclusively shown that the conceptualisation of MEAs as treaties with no existence in their own right has to be reconsidered. MEAs should rather be understood as a specific type of more informal IO, which stands separately and can exert significant influence on their member states.

Second, MEAs can involve states in dynamic processes which expand their authority and create new obligations that were not foreseen when the original treaty was adopted without the adoption of a formal amendment. A core issue that leads to the adoption of expansions in the form of decisions over formal amendments is the ratification process, which would allow states to delay or avoid altogether being bound by new obligations where collective solutions are needed.

Third, the main driving force of such expansive processes is functional pressure. Gaps and ambiguities in the skeletal founding treaties of MEAs can create a functional need for a collective solution which entails the performance of an additional governance function by the MEA, thus expanding the authority it exerts over its member states, analogous to a functional spill-over in European integration. Central areas where this occurs seem to be the need for global participation and the need to ensure compliance with the rules of the MEA to make it function effectively.

Fourth, authority expansions can also happen in the absence of functional pressure, although this second possibility seems less common. When several favourable circumstances combine, a group of states inside an MEA can employ gaps and ambiguities in the treaty to push for an authority expansion that is not functionally necessary but creates additional benefits beyond the functionality of the MEA for this group. This is possible where the general logic of the institution plays in favour of the pushing group and where they can build alliances and broker package deals.

Fifth, a central component to the authority expansions analysed here seems to be the division between developing and industrialised country parties. Both differ significantly in their responsibility for causing environmental harm as well as their capability to undertake protective measures and thus also their obligations under the treaties. But, because of the collective nature of ozone and climate protection, they both need to play their part to solve these issues together. This aspect is shared by all six cases across both analysed MEAs, and both driving forces. Further research is needed to uncover if this aspect can be generalised more broadly.

Thus, *finally*, the results of this investigation can be summarised as follows: MEAs are separate entities in their own right, exercising authority in the international landscape. Their authority can expand beyond the control of their member states without formal delegation. These expansions are predominantly driven by functional pressure, although actor-driven expansions can occur under favourable circumstances. Both driving forces seem to relate to a division between two groups with widely differing capacities.

These results significantly advance our theoretical understanding of informal processes of institutional expansion generally and global environmental politics specifically as one of the most pressing issues of our time. They expand our understanding of MEAs as a type of IO and their dynamic evolvement over time. We begin to recognise MEAs as separate entities in the global landscape, where they used to be understood as mere treaties with no existence of their own. This further broadens the spectrum of existing theories on IO expansion, like the principal-agent-approach and theories of European Integration, to include the investigation into more informal ways of similar processes. This ultimately leads to the question of whether MEAs could even be recognised as actors, similar to how some have begun to recognise full-fledged IOs as such. While answering this question is beyond the scope of the analysis undertaken here, some have begun to investigate this question based on a perspective that corresponds to the one employed here (Gehring and Urbanski 2023; Gehring and Spielmann 2023). Further, my results provide relevant input into recent debates in adjoining areas. This includes the more general debate on authority in global governance as well as the recently emerging investigation into the true capacities of the small-scale bureaucracies of MEAs. It also includes more general discussions on formal institutional adaptation and informal governance in the broader institutional setting, where similar forces could be at play.

This analysis has further provided valuable empirical insights into the inner workings of MEAs and shed much-needed light on the reasons for big institutional developments which set precedents for future MEA developments and significantly changed state commitments. For example, the analysis of the creation of the Multilateral Fund of the Montreal Protocol does not only help to understand this specific case but also expands our understanding of financial arrangements in the realm of international environmental governance more broadly, in particular, the peculiar position of the GEF in relation to MEA-specific funds. Other cases, like the GCF, have found much attention in analysing its institutional structures and its role in addressing climate protection more effectively. The specific circumstances that explain why it was created in the first place, however, had been neglected. Finally, particularly puzzling cases like the definition of developing countries under the Montreal Protocol only become visible once we specifically look for them.

In the end, states becoming parties to MEAs should not be surprised to find out that they might have given up more control than they expected. However, since MEAs are designed to expand over time to fit newly emerging scientific and technological knowledge and changing political circumstances, these expansions should not only be seen as a loss of control but also as opportunities to allow MEAs to adapt to changing circumstances in the most productive way. They are a way for MEAs to evolve to more properly fulfil the task they were created for. Understanding the inner workings of MEAs more accurately will help to steer these processes effectively. Recognising the driving force of an informal expansion under discussion can help to view options more clearly and reduce controversy. Increasing efficiency in this way can be an important factor in reviving MEAs as central frames of global environmental governance and points of reference for the many smaller public and private environmental initiatives. In times when MEAs are known more for their failures than their successes, improving their operation is central to creating a pathway toward a more environmentally conscious future.

It remains to be reiterated that the phenomenon of MEA authority expansion has, so far, rarely been studied. No theoretical or methodological approaches were previously available which could specifically and accurately grasp this phenomenon. As such, the investigation undertaken in this book cannot be more than a first step toward a more extensive research agenda. Now that a starting point has been built, further research is necessary to refine and advance the results presented here. More research is needed to include a larger number of cases under MEAs of different kinds into the analysis and draw on a broader set of differently structured environmental problems to refine generalisations about driving forces of authority expansions and possibly cluster them into different groups according to different types of environmental problems. Ozone and climate protection function in a similar way as they both address the protection of a collective environmental good – the atmosphere – by reducing emissions of specific substances into the air. This might require collective solutions in a different way than those that do not share this same specific nature. For example, while still protecting the environment as a collective good, MEAs like the Basel Convention on the Control of Transboundary Movements of Hazardous Wases and Their Disposals or the Convention on International Trade in Endangered Species of Wild Fauna and Flora (CITES) are focused, instead of emissions into the air, on the movement of things from one place to another and thus can be pinned down to specific states or regions in a different way than it is possible under Montreal or Kyoto. Further, the centrality of the divisions between the developing and industrialised world might not be the same under each MEA. For example, the products that fall under CITES are not as closely linked to development as were chlorofluorocarbons and are greenhouse gases now. Based on the findings of the analysis undertaken here, a structured analysis of a larger number of cases in broader contexts becomes possible. This will also allow to investigate if additional driving forces can be uncovered which have not been at play in the cases analysed here. Further, the investigation of additional cases in different contexts is necessary to uncover if the relevance of the divide between the developing and industrialised world uncovered here can be generalised, or if this aspect can even be abstracted to a more generalised enabling factor such as a differentiation into groups with different levels of capacities or obligations to set about an environmental issue. Based on the results obtained here, it will also become possible to analyse negative cases in which a situation was favourable for an informal authority expansion but it nevertheless did not happen. In this way, specific conditions that need to be present for an informal expansion to take place, can be identified. Further, as mentioned above, the study of MEAs not only as separate entities but as full-fledged actors in their own right needs further advancement which can draw on the results presented here. Finally, the applicability of the concepts uncovered here should be investigated beyond MEAs alone. It is likely that similar processes take place in other types of more informal international institutions. Thus, looking at puzzling developments in these settings through the framework established here might shed more light on the driving forces pushing them on.

A Look into the Future

What remains is a tentative look into the future of MEAs and global environmental protection more generally. In the light of environmental destruction caused by human activity becoming more and more apparent and a growing sense of dread in relation to our future on this planet shared by many, the attention to MEAs is also growing. However, this has also increased awareness of how they seem to fall short of providing the necessary level of protection to keep civilisation as we know it thriving in the long run. Funding in particular remains one of the most central points of contention, not merely in the cases discussed in this book but also at the various COPs taking place as this book is being finished. Further, trust in large-scale international cooperation is ceding, not only in the realm of environmental politics. This is not least because of the tumultuous developments in world politics in recent years which have upstaged the dangers of environmental destruction and uncontrolled climate change and fuel increasing distrust on the national and international level.

The collective, large-scale nature of MEAs can not only lead to the expansions analysed here, but it can also work in the opposite way, leaving states stuck on the lowest common denominator on many issues. A prominent example of this is the adoption at COP 26 in Glasgow of a "phase-down" of coal power instead of the "phase-out" aimed for by many states and NGOs alike (UN 2021). This has given rise to many smaller-scale initiatives and growing private-sector involvement. Now that the initial enthusiasm about the Paris Agreement has faded, the way forward in climate protection seems to be the "climate club", an intergovernmental forum consisting of a smaller number of states who committed themselves to higher ambition in the decarbonisation of key industry sectors, initiated by the G7 in 2022 and launched at COP 28 in 2023. Interestingly, while closely connected to the UNFCCC, the climate club is hosted by the OECD and the International Energy Agency, thus giving it a more Western touch than the globally oriented UN. This development is, however, unlikely to mean the demise of MEAs as such. While they might lose some of their centrality as the main focal points of environmental protection globally, they will certainly remain essential points of orientation. The metaphor of a "coral reef" has been used to describe the function of MEAs as a central hub which attracts all sorts of actors which can organise under its umbrella (Green 2013a, p. 2).

Hope remains that, against all odds, this combination of large-scale cooperation at the MEA level and new smaller-scale initiatives will create more effective advancements in the area of environmental and, in particular, climate protection to finally make the desperately needed advancements to save our planet before it is too late.

References

Barnett, Michael N.; Finnemore, Martha (1999): The Politics, Power, and Pathologies of International Organizations. *International Organization* 53 (4), pp. 699–732. DOI: 10.1162/002081899551048

Bauer, Steffen (2006): Does Bureaucracy Really Matter? The Authority of Intergovernmental Treaty Secretariats in Global Environmental Politics. *Global Environmental Politics* 6 (1), pp. 23–49. DOI: 10.1162/glep.2006.6.1.23

Brunnée, Jutta (2002): COPing with Consent. Law-Making Under Multilateral Environmental Agreements. *Leiden Journal of International Law* 15 (1), pp. 1–52. DOI: 10.1017/S0922156502000018

Churchill, Robin R.; Ulfstein, Geir (2000): Autonomous Institutional Arrangements in Multilateral Environmental Agreements. A Little-Noticed Phenomenon in International Law. *The American Journal of International Law* 94 (4), p. 623. DOI: 10.2307/2589775

Cooper, Scott; Hawkins, Darren; Jacoby, Wade; Nielson, Daniel (2008): Yielding Sovereignty to International Institutions. Bringing System Structure Back In. *International Studies Review* 10 (3), pp. 501–524. DOI: 10.1111/j.1468-2486.2008.00802.x

Gehring, Thomas; Spielmann, Linda (2023): The Treaty Management Organization Established under the UNFCCC and the Paris Agreement: An International Actor in Its Own Right? *International Environmental Agreements* 23, pp. 235–252. DOI: 10.1007/s10784-023-09611-z

Gehring, Thomas; Urbanski, Kevin (2023): Member-Dominated International Organizations as Actors: A Bottom-Up Theory of Corporate Agency. *International Theory* 15 (1), pp. 129–153. DOI: 10.1017/S1752971922000069

Green, Jessica F. (2013a): Order Out of Chaos. Public and Private Rules for Managing Carbon. *Global Environmental Politics* 13 (2), pp. 1–25. DOI: 10.1162/GLEP_a_00164

Haas, Ernst B. (1958): *The Uniting of Europe. Political, Social, and Economical Forces: 1950–1957*. Notre Dame, IN: University of Notre Dame Press.

Heldt, Eugénia; Schmidtke, Henning (2017): Measuring the Empowerment of International Organizations. The Evolution of Financial and Staff Capabilities. *Global policy* 8 (5), pp. 51–61. DOI: 10.1111/1758-5899.12449

Keohane, Robert O. (1989): *International Institutions and State Power. Essays in International Relations Theory*. Boulder, CO: Westview Press.

Krasner, Stephen D. (1982): Regimes and the Limits of Realism. Regimes as Autonomous Variables. *International Organization* 36 (2), pp. 497–510. DOI: 10.1017/S0020818300019032

Kummer, Katharina (1998): The Basel Convention: Ten Years On. *Review of European Community & International Environmental Law* 7 (3), pp. 227–236. DOI: 10.1111/1467-9388.00154

Lake, David A. (2010): Rightful Rules. Authority, Order, and the Foundations of Global Governance. *International Studies Quarterly* 54 (3), pp. 587–613. DOI: 10.1111/j.1468-2478.2010.00601.x

Lindberg, Leon N. (1963): *The Political Dynamics of European Economic Integration*. Stanford, CA: Stanford University Press.

Mahoney, James; Thelen, Kathleen (2009): *Explaining Institutional Change*. Cambridge: Cambridge University Press.

Niemann, Arne (2006): *Explaining Decisions in the European Union*. Cambridge: Cambridge University Press.

Tranholm-Mikkelsen, Jeppe (1991): Neo-Functionalism. Obstinate or Obsolete? A Reappraisal in the Light of the New Dynamism of the EC. *Millennium* 20 (1), pp. 1–22. DOI: 10.1177/03058298910200010201

UN (2021): *COP 26: Together for Our Planet*. Available online at www.un.org/en/climatechange/cop26, checked on 10/22/2023.

Wiersema, Annecoos (2009): The New International Law-Makers? Conferences of the Parties to Multilateral Environmental Agreements. *Michigan Journal of International Law* 31 (1), pp. 231–287.

Index

Note: Page locators in **bold** and *italics* represents tables and figures, respectively.

For Product Safety Concerns and Information please contact our EU representative GPSR@taylorandfrancis.com Taylor & Francis Verlag GmbH, Kaufingerstraße 24, 80331 München, Germany

Batch number: 10397794

Printed by Printforce, the Netherlands